PROGRESSES IN CHEMICAL SENSOR

Edited by **Wen Wang**

Progresses in Chemical Sensor
http://dx.doi.org/10.5772/61376
Edited by Wen Wang

Contributors

Saeed Olyaee, Erica Perry Murray, Ling Cui, Sukumar Basu, Morteza Enhessari, Ali Salehabadi, Keshavaditya Golla, Ronald Coutu, Nathan Glauvitz, Ivan Medvedev, Douglas Petkie, Mohd Nizar Hamidon, Wen Wang

Published by InTech

Janeza Trdine 9, 51000 Rijeka, Croatia

Publishing Process Manager Romina Rovan

Technical Editor SPi Global

Cover InTech Design team

First published August, 2016

Printed in Croatia

Additional hard copies can be obtained from orders@intechopen.com

Progresses in Chemical Sensor, Edited by Wen Wang

p. cm.

Print ISBN 978-953-51-2532-7
Online ISBN 978-953-51-2533-4

For EU product safety concerns:
IN TECH d.o.o., Prolaz Marije Krucifikse Kozulić 3, 51000 Rijeka, Croatia,
info@intechopen.com or visit our website at intechopen.com.

Meet the editor

Wen Wang, a full-time professor at the Institute of Acoustics, Chinese Academy of Sciences (IOA-CAS), received his MS degree from the Central South University of China in 2002 and his PhD degree from CAS in 2005. From 2005 to 2009, he worked as a postdoctoral researcher and research assistant professor at the Microsystem Laboratory of Ajou University in South Korea. In the two following years, he worked at the Freiburg University in Germany as a guest professor supported by the Humboldt Foundation and rewarded as "experienced researcher" by the Humboldt Foundation for his outstanding research work in SAW sensor technology. In 2011, he was elected to "Hundred-talent program" of the CAS and then named as the CAS specially appointed researcher. His current research involves SAW devices for physical and chemical sensor applications and wireless sensors. Prof. Wang is the author or coauthor of more than 100 papers in a refereed international journal, more than 60 communications in international conferences, over 20 patents, and 2 books.

Contents

Preface

Although the history of chemical sensor dates back not long ago, it has attracted great research interest owing to its many excellent properties such as small size, satisfactory sensitivity, larger dynamic range, low cost, and easy-to-realize automatic measurement and online or in situ and continuous detection. With decades of vigorous research works, various sophisticated chemical sensors have been widely used in environmental conservation and monitoring, industrial process monitoring, gas composition analysis, medicine, national defense and public security, and on-site emergency disposal. The chemical sensor is the analyzer that responds to a particular analyte in a selective and reversible way and transforms input chemical information, ranging from the concentration of a specific sample component to total composition analysis, into an analytically electrical signal. Obviously, the sensor performances are determined mainly by the characteristics of the sensing materials and mechanism therefrom; hence, advanced sensing materials and new response mechanism become the focus of research in chemical sensor and are reflected in numerous literatures.

This book is an attempt to highlight the progresses in functional structures, sensing materials, and response mechanism of chemical sensors. It is composed of eight chapters and divided into five sections categorized by the working principle of the chemical sensor. The sensor response mechanism, sensing structure and materials, and future trend of the chemical sensors were reviewed in details. Section 1 is the introductory of this book, which introduces the conceptual framework, main types, and development trend of chemical sensors. The next section provides an introduction to the latest research progress relating to the optical chemical sensors using the principle of photo-acoustic and photonic crystal. Section 3 reviews some research achievements of metal oxide and semiconductor-based chemical sensor by using some new sensing materials as graphene and perovskites, and section 4 describes advances in electrochemical chemical sensor for sensing nitric oxide exhaust gas. Section 5 provides the introduction to surface acoustic wave (SAW) sensor. As a relatively new chemical sensor, it is favored and very promising for sensing trace poisonous gas owing to their small size, low cost, high sensitivity and reliability. The sensor was composed of SAW sensor chips and selective sensing materials deposited along the SAW propagation path. Obviously, the accurate identification of the target species mainly depended on the sensing materials, which are the main topics in this section.

It is my pleasure that this collection of the up-to-date information and latest research progress on chemical sensor in this book will be of great interest to all those working on chemical sensors. I would like to acknowledge the hard work and dedication of all the contributing authors. Also, I am particularly grateful to Ms. Romina Rovan, the publishing process manager of this book, for her great help in proposal collection, evaluation, and manuscript editing.

Wang Wen PhD
Professor
Institute of Acoustics, Chinese Academy of Sciences
Beijing, China, PR

Section 1

Introductory Chapter

Chapter 1

Introductory Chapter: What is Chemical Sensor?

Wang Wen

Additional information is available at the end of the chapter

http://dx.doi.org/10.5772/64626

1. Introduction

The chemical sensor is an analyzer that responds to a particular analyte in a selective and reversible way and transforms input chemical quantity, ranging from the concentration of a specific sample component to a total composition analysis, into an analytically electrical signal [1, 2], as depicted in **Figure 1**. The chemical information may originate from a chemical reaction by a biomaterial, chemical compound, or a combination of both attached onto the surface of a physical transducer toward the analyte. The chemical sensor subject is an emerging discipline formed by the multidisciplinary study among chemistry, biology, electricity, optics, mechanics, acoustics, thermology, semiconductor technology, microelectronics technology, and membrane technology.

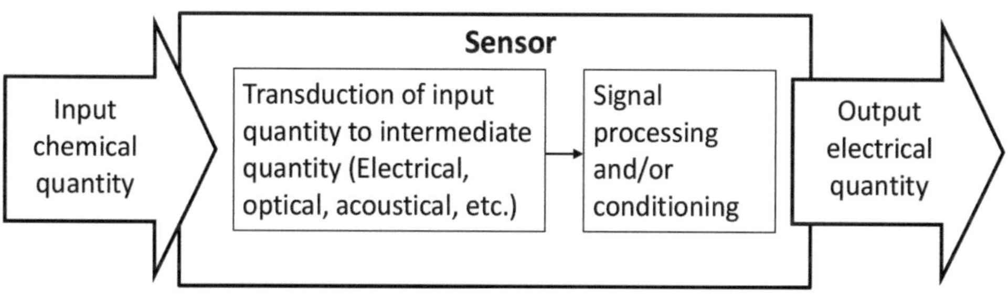

Figure 1. Sensor principle. Schematic diagram of a sensor that produces an electrical output in response to the presence of an input quantity.

Although the history of chemical sensor dates back not long ago, it has gained increasing attraction for applications in environmental monitoring, industrial process monitoring, gas composition analysis, medicine, national defense and public security, and on-site emergency disposal owing to its many excellent properties such as small size, satisfactory sensitivity, larger dynamic range, low cost, and easy to realize automatic measurement and online or in situ and continuous detection. Hence, the chemical sensor becomes one of the most active and effective directions of modern sensor technology.

2. Classification of chemical sensor

According to the working principle, the chemical sensor can be classified into many types such as optical, electrochemical, mass, magnetic, and thermal. The optical chemical sensor is based on the changes in optical phenomena analysis arising from the interaction between the analyte and the receiver. The electrochemical sensor utilizes electrochemical effect among the analytes and featured electrodes. The working principle of the mass sensor depends on the quality change induced by the mass loading from the adsorption toward the analyte by the special modification of sensor surface. The magnetic device is based on the magnetic properties in analyte adsorption, whereas the thermal sensor utilizes the thermal effect generated by the specific chemical reaction or adsorption process.

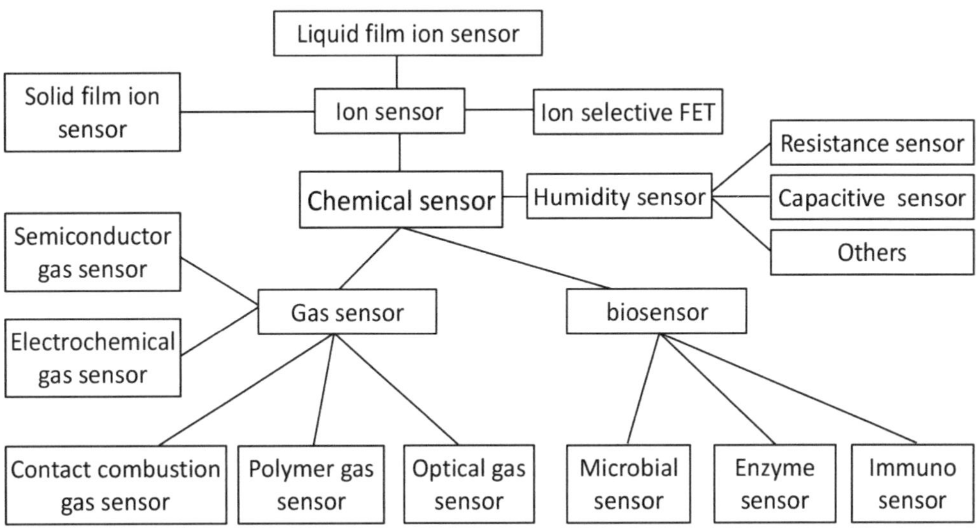

Figure 2. Classification of chemical sensors based on sensing objects.

Another way to categorize the chemical sensors is based on the object to be detected, that is, the chemical sensors can be classified as gas sensors for trace gas analysis and monitoring, various ion sensors represented by the pH sensor, humidity sensor, and biosensors made by biological characteristics. **Figure 2** lists the main categories of chemical sensors toward various objects.

It is well known that the environment monitoring is the major application field of the chemical sensor, therein, the gas sensor plays a pivotal role. The key performance indicators of the gas sensor include sensitivity, selectivity, and stability, which are determined mainly by the characteristics of the sensing material and mechanism. By utilizing the advanced materials and the new response mechanism, the gas sensitive characteristic of the sensor can be improved significantly. Generally speaking, the gas sensor is categorized mainly by its working principle as a semiconductor type, an electrochemical type, a solid electrolyte type, a contact combustion type, a photochemical type, and a polymer type. Each type is introduced in the paragraphs below.

2.1. Semiconductor gas sensor

Semiconductor gas sensor is a sensitive element made by metal oxides or metal semiconductor oxide materials [3]. As regards the electrical conductivity sensors, the resistance of their active sensing layer changes due to contact with the gas to be detected. Since the first semiconductor metal-oxide-ceramic gas sensor was reported in 1962, the semiconductor gas sensors have become the most comprehensive and widely used gas sensors.

2.2. Electrochemical gas sensor

The electrochemical gas sensor can be categorized into galvanic cell type, controlled potential electrolysis type, coulometric type, and ion-selective electrode type [4]. The galvanic cell gas sensor evaluates the target gas composition by measuring the shift in current. The controlled potential electrolysis gas sensor senses the target gas by measuring the electrolytic current and is different from the galvanic cell sensor, and a specific voltage should be imposed externally. Furthermore, the oxygen in blood can also be detected in addition to CO, NO, NO_2, and SO_2. The coulometric gas sensor detects the target species by measuring the current generated by the interaction between the gas and the electrolyte. The ion-selective electrode gas sensor has appeared earlier, and it detects the gas by measuring the ion current with high sensitivity and excellent selectivity.

2.3. Contact combustion gas sensor

The contact combustion gas sensor includes direct contact sensor with the combustion and catalysis combustion sensor [5]. Its working principle is that oxidative combustion of the combustible gas occurs directly or by using catalysts in the energized state, accordingly, the burning gas sensitive material (Pt wire) is heated up, resulting in the shift in the resistance value. By evaluating the resistance shift, the gas concentration can be extracted. Sometimes, such a sensor is called a thermal conductivity sensor, which is widely used for sensing combustible gas in petroleum chemical plant, shipyards, mine tunnels, kitchens, and bathrooms.

2.4. Optical gas sensor

Optical gas sensors include infrared absorption sensors, spectrum absorptive sensors, fluorescence sensors, and fiber sensors [6], in which, the infrared absorption sensor is the most widely used for sensing gas by measuring and analyzing the infrared absorption peak from various gas adsorption. Such a sensor has a lot of outstanding advantages of excellent antivibration and antipollution ability, autocorrection, and possibility of continuous and longer dynamic monitoring.

2.5. Polymer gas sensor

The polymer gas sensitive materials have grown enormously in the past several years, which plays an important role in sensing trace poisonous gas because of its easy operation, simple process, good selectivity at normal temperature, low price, and easy to combine with the micro structure or surface acoustic wave (SAW) devices [7].

3. Development trend of chemical sensors

Environmental protection and monitoring, prevention and treatment of disease and the continuous improvement of people's quality of life are still the main application areas of the chemical sensors in the foreseeable future. Nowadays, with new challenges as well as new opportunities, the features of the chemical sensors trend are as follows:

1. A variety of gas sensors used widely in environmental protection and monitoring are highly valued, which are the mainstays of three mainstreams of chemical sensors. High sensitivity, miniaturization, integration, and low cost are still the development tendency of the gas sensors.

2. Biosensor ranks second in three mainstream of current chemical sensors. The twenty-first century is the century of biology. As an important part of biology, the biosensor provides the important support for the biological technology. It is not only an important aspect of the international competition of high-tech but also a window to evaluate the biological level of a country. Nowadays, the research on biosensor is strengthened in many countries to adapt to rapid development of biology. Among them, the microarray technology dominated by the DNA or cDNA biochip is one of the emphases for biosensor research. Nanobiosensing technology formed by a combination of biosensing technology and nanotechnology would be a new growth point of the biosensors, and a unprecedented new situation will be created based on the new applications of the silicon and nonsilicon-based micro biosensor in biomedical, environmental monitoring and instrumentation industries.

3. Electrochemical sensor ranks only the third in the mainstream of current chemical sensors, but it is very active. The electromechanical sensor dominated by ion selective electrode (ISE) will continue to trend to high sensitivity, low detection limit, fast response, and long-operating life. The appearance of the new materials such as organometallic compound, organometallic polymer, and macrocyclic compounds and their complexes represented by the metal porphyrin (metalloporphyrins) offers a solid foundation for gradual perfection of all solid-state ion-selective electrode, electrochemiluminescence (ECL), and photoelectrochemical sensor. Some new technologies such as nanoparticle LB films, molecular imprinting technology (MIT), and nanoelectrode arrays, and appearance of

scanning electrochemical microscopy (SECM) and electrochemical impedance spectroscopy (EIS) will make the electrochemical sensor vigorous and exuberant and expand some new applications in environmental protection, food industry, biology, and medicine.

4. A new generation of bionic sensors, such as electronic nose and electronic tongue, will be a new growth point of chemical sensor in the coming century. It is an important aspect of artificial intelligence study to simulate the function of human bodies and facial features (five human senses) to recognize various substances and environments effectively. Since the breakthrough in taste sensor was obtained in the 1990s, the substances with five major tastes of sweetness, sourness, bitterness, spiciness, and saltiness can be detected quantitatively by using the multichannel technology. Further improvement of the taste sensor in selectivity, repeatability, and durability will lead to more applications in the food industry to improve the quality and quality control of beverages, fruits, and vegetables. Electronic nose is a multichannel gas sensor array that provides an effective way to detect toxic gas, explosive gas, as well as the released smell of drugs, explosives, and food.

5. The emergence of the microchemical sensor and its array is an important sign of the chemical sensor with excellent performance entering the new century. With the development of solid-state and microelectronic technology and the appearance of the MEMS sensors aiming for high precision, low driving, low power consumption, small size and fast response, the chemical sensor will be turned over a new leaf, and enters into an unprecedented "Micro world."

Author details

Wang Wen

Address all correspondence to: wangwenwq@hotmail.com

Institute of Acoustics, Chinese Academy of Sciences, Beijing, China

References

[1] Janata. Introduction: modern topics in chemical sensing, Chemical Reviews, 2008, 108(2): 327–328.

[2] Catterall, R.W. Chemical sensors. Oxford University Press, Oxford, UK, 1997: 4–30

[3] Yong, S.K., Ha, S.C., Kim, K., Yang, H. Room-temperature semiconductor gas sensor based on nonstoichiometric tungsten oxide nanorod film, Applied Physics Letters, 2005, 86(21):213105.

[4] Ishiji, T., Imaya, H. Advanced electrochemical sensor for gas detection, IEEJ Transactions on Sensors and Micromachines, 2015, 135(8):276–280.

[5] Takashi, O., Yoshiaki, I., Kazuhiro, T. Detection of decomposed compounds from an early stage fire by an adsorption combustion-type sensor. Sensors and Actuators B: Chemical, 2005, 108(1–2), 473-477.

[6] Airoudj, A., Debarnot, D. Design and sensing properties of an integrated optical gas sensor based on a multilayer structure, Analytical Chemistry, 2008, 80(23):9188–94.

[7] Wang, W., He, S., Li, S., Liu, M., Pan Y., Enhanced sensitivity of SAW gas sensor coated molecularly imprinted polymer incorporating high frequency stability oscillator, Sensors and Actuators B: Chemical, 2007, 125: 422–427.

Section 2

Optical Chemical Sensor

Chapter 2

MEMS-Based Terahertz Photoacoustic Chemical Sensing System

Nathan Glauvitz, Ronald A. Coutu,
Ivan R. Medvedev and Douglas T. Petkie

Additional information is available at the end of the chapter

http://dx.doi.org/10.5772/62571

Abstract

Advancements in microelectromechanical system (MEMS) technology over the last several decades has been a driving force behind miniaturizing and improving sensor designs. In this work, a specialized cantilever pressure sensor was designed, modeled, and fabricated to investigate the photoacoustic (PA) response of gases to terahertz (THz) radiation under low-vacuum conditions associated with high-resolution spectroscopy. Microfabricated cantilever devices made using silicon-on-insulator (SOI) wafers were tested in a custom-built test chamber in this first ever demonstration of a cantilever-based PA chemical sensor and spectroscopy system in the THz frequency regime. The THz radiation source was amplitude modulated to excite acoustic waves in the chamber, and PA molecular spectroscopy of a gas species was performed. An optical measurement technique was used to evaluate the PA effect on the cantilever sensor; a laser beam was reflected off the cantilever tip and through an iris to a photodiode. As the cantilever movement deflected the laser beam, the beam was clipped by an iris and generated the PA signal. Experimental data indicated a predominantly linear response in signal amplitude from the photodiode measurement technique, which directly correlated to measured cantilever deflections. Using the custom-designed PA chamber and MEMS cantilever sensor, excellent low-pressure PA spectral data of methyl cyanide (CH_3CN) at 2 to 40 mTorr range has been obtained. At low chamber pressures, the sensitivity of our system was 1.97×10^{-5} cm^{-1} and had an excellent normalized noise equivalent absorption (NNEA) coefficient of 1.39×10^{-9} cm^{-1} W Hz$^{-1/2}$ using a 0.5 s signal averaging time.

Keywords: cantilever, MEMS, photoacoustics, spectroscopy, terahertz sensor

1. Introduction

Chemical sensing and the spectroscopic detection of chemical species can be made through various means. In this work, a direct sensing technique was performed by measuring the photoacoustic (PA) response of a microelectromechanical system (MEMS) cantilever when a gaseous sample was excited with a terahertz (THz) radiation source. The "energy deposited in the sample is measured directly" [1] through the PA effect and is proportional to the radiation source power. The PA effect was first published by Alexander Graham Bell in 1880, when he found that modulated sunlight incident on a thin disk generated sound waves [2]. Since its discovery, the PA effect has been used to study solid, liquid, and gaseous states of matter. Traditional molecular absorption spectroscopy systems measure the electromagnetic energy that is transmitted through a sample species. A disadvantage of the absorption spectroscopy technique is that it requires a baseline measurement, which is needed to account for standing waves in the system that may obscure absorption lines. In this research, the MEMS cantilever PA response to modulated THz radiation was used to investigate chemical sensing applications and collect molecular absorption spectra. An advantage over the traditional absorption spectroscopy technique is that the PA technique used here does not require a baseline measurement and it is not sensitive to baseline variations seen in traditional absorption spectroscopy techniques.

Molecules can absorb and dissipate energy through different pathways such as electronic, radiative, vibrational, and rotational pathways or through translational means via collisions [3, 4]. In recent years, many techniques have been employed for the detection of trace gases and molecular spectroscopy [1]. The PA effect is the result of molecules absorbing energy from an electromagnetic wave, where the absorbed energy is then released through collisions and other nonradiative pathways into translational energy, resulting in an increased pressure. This pressure wave can then be detected by a microphone or other pressure-sensitive device. Several different sensor methods to pick up PA waves have been implemented over the years; among them are the tuning fork [5–8], membrane microphone [9, 10], bridge [11], and cantilever designs [11–20]. As is true for any sensor configuration, the sensor design must be optimized to achieve maximum sensitivity to the anticipated changes in pressure caused by the excitation radiation captured in the PA cell.

Additional applications of MEMS and THz radiation technology have also been investigated for imaging purposes. The incorporation of superconductors [21] and metamaterials [22] on MEMS devices has enabled the direct absorption and sensing of THz radiation. As acoustic sensors, arrays of piezoelectric cantilever and spiral beam-supported diaphragm transducers have shown increased sensitivity by approximately 30 times when many transducers were connected in parallel [23]. Along those same lines, arrays of piezoelectric cantilevers have the potential to be used for THz PA imaging applications.

1.1. PA performance figures of merit

In PA spectroscopy and trace chemical detection systems, there are several figures of merit that are traditionally used to evaluate system performance. The first metric that must be determined is the signal-to-noise ratio (SNR). The SNR of the system is

$$SNR = \frac{PA_{Signal}}{PA_{Noise}} \tag{1}$$

where the PA_{Signal} is measured on an absorption line and the PA_{Noise} is the PA signal level evaluated in a region away from absorption lines. A high SNR is desirable because it plays a role in calculating another performance metric of the system, the sensitivity, which is described as what minimum absorption strength the system can detect. The sensitivity α_{min} of a system is

$$\alpha_{\min} = \frac{\alpha_{peak}}{\mathrm{SNR}} \tag{2}$$

where α_{peak} is the strength of the measured absorption coefficient. Because PA systems can be configured differently, another useful figure of merit is the normalized noise equivalent absorption (NNEA) coefficient and it is expressed as

$$\mathrm{NNEA} = \alpha_{\min} P_o \sqrt{T}. \tag{3}$$

NNEA calculations allow for a better performance comparison between different PA systems by taking into account the system sensitivity, the radiation source power P_o, and the PA signal averaging time T.

1.2. PA detection systems

The PA excitation radiation source has to be specifically selected for the absorption lines of the chemical or chemicals under investigation. Radiation sources that have been commonly used are black-body sources, arc lamps, microwaves, LEDs, lasers, or tunable lasers. Kuusela et al. [12] used three different LED radiation sources, centered at 3.4, 4.2, and 7.0 μm wavelengths to excite the gaseous specimens to perform trace gas detection. Their experiment tested methane, propane, CO_2, and sulfur dioxide gases using a 3 × 1.5 × 0.01 mm^3 (length × width × thickness) silicon cantilever with a 5 μm gap around the beam. The small sensitive cantilever PA gas sensor design used a laser interferometer to measure cantilever displacement. A figure of merit for performance of a trace gas detection system can be given by the lowest achievable detection limit, which was 6 ppm for propane with a 1 s sample integration time. Experiments were performed at atmospheric pressure (760 Torr) with the gas species diluted with nitrogen to control the concentration [12].

Ledermann et al. [11] constructed a miniaturized PA cell with dimensions of approximately 15 mm in length × 8 mm in diameter to test piezoelectric acoustic sensors configured in bridge and cantilever structures for detecting CO_2 in a nitrogen-diluted mixture at atmospheric pressure. Cantilever devices were approximately 2 mm long × 2 mm wide and were fabricated using a silicon wafer that was backside etched to determine the device layer thicknesses that ranged from 5 to 20 μm. A pulsed incandescent lamp and an interference IR filter were used in the test chamber to produce 4.25 μm wavelength radiation to excite the CO_2 gas mixture. In the PA tests, depending on the CO_2 concentration, they measured 200 to 1300 μV from the piezoelectric layer on the devices, leading to a minimum sensitivity of 330 ppm CO_2 [11].

In an earlier work by Kauppinen et al. [18], a more flexible 4 × 2 × 0.005 mm^3 silicon cantilever with a 30 μm gap was used to detect methane down to 0.2 to 0.8 ppb with their PA sensor and chamber configuration. An optically chopped black-body source and a band-pass filter allowed 2.94 to 3.85 μm wavelength radiation to excite the methane gas, which has an absorption line at 3.33 μm. The methane gas tested was also diluted with nitrogen and the PA measurements were performed at a reduced pressure of 465 Torr [18].

McNaghten et al. [14] reported on an interferometer-based PA cantilever detection method that used four multiplexed tunable diode lasers at wavelengths of 1.534, 1.567, 1.568, and 1.62 μm as the radiation sources to investigate acetylene, carbon monoxide, CO_2, and methane gases in nitrogen dilution. PA measurements using a 6 × 1.5 × 0.01 mm^3 cantilever were taken on the gas samples over a range of pressures, but they found that the optimal PA signal for their system configuration was 93 Torr. Nominal data collection duration used to analyze a gaseous sample was approximately 2.62 s. To achieve the best sensitivity, signals were averaged over several minutes. The lowest NNEA for their system was achieved with acetylene gas at 1.4×10^{-9} cm^{-1} W $Hz^{-1/2}$. They evaluated the trace detection system performance over a long period of time and found that the Michelson interferometer set-up showed some instability in signal intensity over a 12 h data collection period. The fluctuations in signal intensity were attributed to acoustic noise in the room and phase shifts in the interferometer set-up seen at the photodiode [14].

Recently, Peltola et al. [17] used an interferometer-based analyzer system that included the cantilever sensor, PA chamber, and software, manufactured by Gasera Ltd. (Turku, Finland), to investigate the detection limits of hydrogen cyanide (HCN) and methane. The cantilever sensor was 5 × 1.5 × 0.01 mm^3, and Peltola et al. used a continuous-wave optical parametric oscillator (OPO) centered at ~3 μm wavelength as the radiation source. The detection limit for HCN was calculated to be 190 ppt in nitrogen and the NNEA was 1.8×10^{-9} cm^{-1} W $Hz^{-1/2}$ with an OPO power of 0.5 W and a 1 s signal averaging time. PA measurements were conducted at 300 Torr (~0.4 atm), whereas the HCN and nitrogen mixture were flowed through the chamber at ~1 L min^{-1} [17]. With another analyzer from Gasera Ltd., Sievilä et al. [19] tested six different cantilever designs in the system for PA detection, testing solid-phase materials of carbon black and polyethene. Their most sensitive cantilever design had dimensions of 5 × 1.2 × 0.005 mm^3 and a SNR = 189 while testing polyethene.

The cantilever PA systems discussed previously performed trace gas analysis to find the minimum detectable concentration of a selected gas or gases in a nitrogen-diluted environment. The PA measurements were conducted at reasonably high chamber pressures ranging from 93 to 760 Torr. Due to the high chamber pressures used, the PA signal was often stronger at modulation frequencies lower than the resonant frequency of the cantilever sensors. For spectroscopy applications in the THz spectral region, the PA measurements should be made at the lowest chamber pressure possible. A low chamber vacuum level minimizes the effects of pressure broadening on the absorption coefficient and allows for closely spaced absorption lines to be resolved. One of the earliest applications of THz PA spectroscopy was done by Krupnov and Burenin in the 1970s, when they constructed a submillimeter spectrometer that used a 3-μm-thick, 5 cm in diameter Mylar capacitive membrane microphone in a PA cell. They used broadband submillimeter backward wave oscillators (BWOs) as the radiation sources, which could produce frequencies of 0.199 to 1.07 THz. The chamber pressures tested ranged from 0.05 to 10 Torr and were able to achieve sensitivities of 6×10^{-9} cm^{-1} with a radiation source power of 10 mW [9].

The THz diode amplifiers used here as a radiation source for PA signal generation are modular; through different combinations of amplifier chains, the system can span 0.250 to 1.5 THz frequency range with great accuracy. The small THz PA spectroscopy system developed in this work has several advantages over traditional THz spectroscopy absorption cells. A major advantage is the reduction in size, as traditional spectroscopy absorption cells can be large, ranging from 3 to 10 ft long to increase sensitivity. Cantilever sensor designs have been shown to be more sensitive than bridge-type microphone designs [11], and the sensitive MEMS cantilever design used here enables the absorption cell size to be greatly reduced. Cantilever deflection caused by a PA pressure load dissipates the energy through bending. For a comparable membrane or bridge sensor, the deflection occurs through stretching and the devices would experience much less deflection for the same applied load conditions.

Categories for chamber pressure operating conditions have classically been divided into intrinsic, molecular, or viscous regimes [24]. Each pressure regime was named after the dominant damping mechanisms. Previously discussed cantilever PA systems were operated at higher chamber pressures of 93 to 760 Torr, which is in the portion of the pressure spectrum dominated by viscous damping effects. The molecular regime has been defined in the range of 7.5 to 750 mTorr (1–100 Pa) [24] and the intrinsic regime is defined as pressure environments less than 7.5 mTorr. A significant contribution of the research presented in this work is in the low 2 to 60 mTorr chamber pressure regimes that were investigated for the PA molecular spectroscopy applications. PA measurements in this work were performed at the lower portion of the molecular regime and just crossing over into the intrinsic damping dominant regime.

The analysis of the cantilever behavior in the PA system is first discussed through analytical equations and then finite element model (FEM) using CoventorWare® to evaluate cantilever performance. An overview of the MEMS device fabrication is then presented, which highlights the methodology used to create the cantilever PA sensor. The experimental PA test set-up is then presented, which used an extremely stable optical measurement technique to observe and record the cantilever PA signal. Cantilever dynamic behavior in the PA chamber and PA spectral data collection techniques for low-pressure environments are then presented. Finally, the sensitivity of the system and the impact of future potential applications of this THz sensing method are discussed.

1.3. PA molecular spectroscopy

In the THz regime, molecules in a low-vacuum environment absorb energy in narrow quantized rotational states. The associated absorption frequencies are a function of the molecular structure, whereas the absorption strength depends on the electric dipole moment and structure. The characteristic absorption line shape coefficients of a molecule is subject to many different sources of line broadening, which can be established through the convolution with an appropriate broadening function. At low pressures, the dominant source of spectral line broadening is Doppler broadening, which can be represented with a Gaussian function. Doppler broadening is due to the thermal motion of the absorbing atoms, which leads to small variations in the absorbed frequency. As the chamber pressure increases, pressure broadening will play more of a role, as the absorbing molecules will collide with neighboring molecules more frequently. Pressure broadening, represented by a Lorentzian function, becomes more dominant at higher pressures. Often, spectral line widths are a function of both Doppler and pressure broadening, which can be represented through a Voigt profile and these line broadening parameters are unique to each molecule. In the THz spectral range, the optimal operating pressure is near ~10 mTorr when the Doppler broadening and pressure broadening contributions to the line shape are nearly equal.

To model the PA performance of the cantilever in the THz PA chamber, the order of magnitude of the pressure change was estimated through the use of the kinetic theory of gases and the ideal gas law. PA chamber designs can take on many different configurations such as cylindrical, Helmholtz [3], or differential [25] PA system designs. For this PA analysis, the absorption cell is assumed to have a cylindrical shape with a characteristic length l and radius r. It is also assumed that the THz radiation beam energy from the source is uniformly distributed and that the radius of the beam closely matches to that of the chamber radius. The radiation source power defined as P_0 is the amount of power inserted into the chamber, whereas P is the amount of unabsorbed power that exits the chamber. Over long distances, the molecular absorption of radiation is best described by Beer's absorption law, and through a short chamber length l, the absorbed power ΔP follows the relationship:

$$\Delta P = P_0 - P = P_o p_x \alpha l \tag{4}$$

where α is the absorption coefficient and p_x is the partial pressure factor of the gas under investigation. The partial pressure factor here describes the volumetric fractional abundance of the species and can be set equal to one, as no backfill of gas is used to dilute the gas species under investigation. **Figure 1** shows a segment of simulated absorption coefficient α (m^{-1}) spectra for methyl cyanide (CH_3CN) at 18 mTorr.

http://dx.doi.org/10.5772/62571

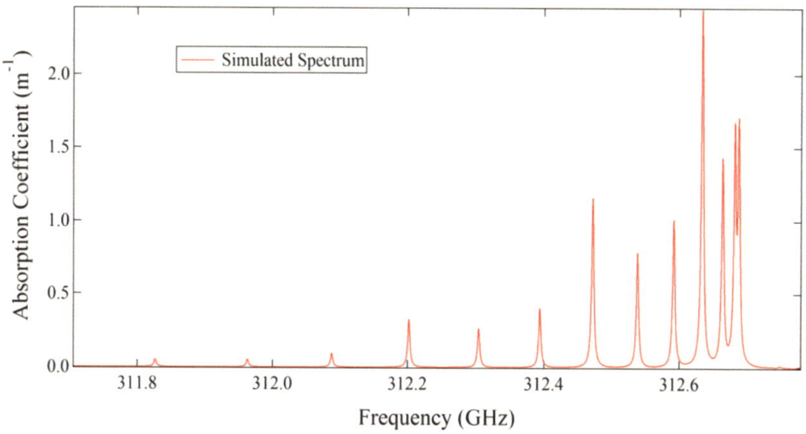

Figure 1. Small region of the simulated absorption coefficient of CH_3CN at 18 mTorr with applied pressure broadening conditions.

Based on the predicted absorption coefficients, an equation for the PA change in pressure Δp can be given as

$$\Delta p = \frac{P_o p_x \alpha l}{3V f_m} = \frac{P_o \alpha}{3\pi r^2 f_m} \tag{5}$$

where the anticipated change in chamber pressure per cycle as a multivariable function of radiation source power, V is the absorption chamber volume, p_x is the partial pressure factor of the gas, α is the absorption coefficient, r is the chamber radius, and f_m is the radiation modulation frequency. Based on these design space parameters, the anticipated changes in pressures could span a large range depending on the gas constituent tested and the vacuum level in the chamber. Under ideal conditions, the change in chamber pressure per cycle is expected to be approximately 18 μTorr when a radiation source power of 0.1 mW, $\alpha = 1\ m^{-1}$, $r = 5$ mm, and $f_m = 550$ Hz are used.

The sensitive microfabricated MEMS cantilever sensors created here were the key technology that significantly reduced the size of the system and enabled the low-pressure spectral measurements. A complete system integration of the sensitive MEMS cantilever sensor and the narrow line, broad spectral capabilities of the THz radiation source used with the custom low-pressure PA chamber create a uniquely powerful spectral chemical analysis tool.

2. Cantilever modeling

Cantilever motion in an under damped system can be described by Newton's second law of motion, where all the internal and external forces acting on the cantilever are set up in a detailed balance equation. The resulting differential equation describes the cantilever position based on the forces involved and the initial conditions of the system. Of particular interest are two cases of motion, driven oscillation and damped harmonic oscillation. Using Newton's second law of motion, the equation for an under damped driven oscillator under an external driving force is represented as

$$m\frac{d^2x}{dt^2}+c\frac{dx}{dt}+kx=F\cos(\omega t+\varphi) \tag{6}$$

where m is the mass of the object, c is the damping coefficient, k is the spring constant, x is the position, dx/dt is the velocity, d^2x/dt is the acceleration, F is the amplitude of the driving force, ω is the excitation angular frequency, and φ is the phase term. A steady-state solution for displacement x for a sinusoidal applied force is

$$x(t)=A\cos(\omega t+\varphi). \tag{7}$$

The amplitude of the displacement A is a function of the modulation frequency ω, the angular resonant frequency ω_o of the cantilever and of the applied force F. When the system is modulated at the resonant frequency ($\omega=\omega_o$), the expression can be simplified to

$$A(\omega)=\frac{F}{m\sqrt{\left(\omega_o^2-\omega^2\right)^2+4\left(\frac{1}{\tau}\right)^2\omega^2}}\cong\frac{F\tau}{2m\omega}=\frac{F}{c\omega} \tag{8}$$

where the time constant τ, defined as $\tau = 2$ m/c, represents how quickly the system dampens or responds to external stimuli. An examination of Eq. (8) reveals that decreased damping and a lower resonant frequency of the cantilever allow for the greatest amount of deflection for a given applied force.

Traditional cantilever design parameters of length, width, and thickness were used to create designs sensitive to the lower end of the anticipated pressure spectrum. As discussed above, cantilever spring constant and cantilever resonant frequencies are of great importance to the design to maximize sensitivity. An equation for the spring constant k of a rectangular cantilever beam [13] is

$$k=\frac{2}{3}E_Y w\left(\frac{h}{L}\right)^3 \tag{9}$$

http://dx.doi.org/10.5772/62571

where E_Y is the Young's modulus of the material, whereas the other parameters are the cantilever dimensions of length L, width w, and thickness h. A cantilever with the above dimensions then has a resonant frequency f_o that is then described as

$$f_o = \frac{1}{2\pi}\sqrt{\frac{k}{m_{eff}}} \tag{10}$$

where m_{eff} is the effective mass of the cantilever. Using Rayleigh's method of the conservation of energy, where ρ is the density of the cantilever material, it can be shown that the cantilever effective mass under a uniform distributed pressure load is

$$m_{eff} = \frac{104}{405}\rho whL. \tag{11}$$

Then, with these anticipated PA pressure conditions estimated through Eq. (5), potential cantilever designs were evaluated in the CoventorWare® FEM software. FEM using modal analysis of cantilever designs was performed in an earlier work and it showed that, as the cantilever width over length ratio increased, the second harmonic torsional mode frequency decreased [26]. To minimize the amount of energy dissipated through higher harmonic modes, it is advantageous to use the smaller width over length ratio cantilever designs. Using the harmonic modal analysis tool, a harmonic pressure load was applied to the surface of the cantilever as a function of frequency to find the maximum magnitude of the cantilever tip deflection. A cantilever damping parameter was used in the FEM software, defined as a fraction of the critical damping coefficient c_0, which is

$$c_0 = 2\sqrt{mk} \tag{12}$$

where m is the mass of the beam and k is the spring constant. For the simulations, the damping conditions for the analysis were set to 0.1% of the critical damping coefficient to simulate a low-pressure environment. **Figure 2** shows the tip displacement amplitude response of a $5 \times 2 \times 0.01$ mm^3 silicon cantilever under four load conditions. Simulated sinusoidal load pressures of 0.03 to 7.5 μTorr resulted in amplitude displacements of 0.14 to 2.8 μm, respectively. The modeled results predict cantilever deflections that could easily be measured optically with a laser interferometer set-up and provided a reasonable estimate of PA performance of the cantilever design.

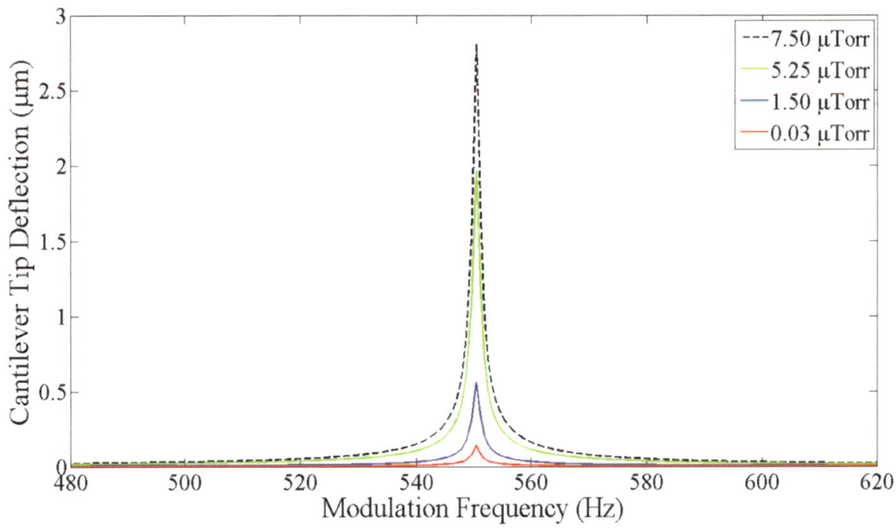

Figure 2. FEM resultant tip deflections from modal harmonic analysis of a 5 × 2 × 0.01 mm^3 silicon cantilever design under four periodic load conditions using 0.1% of the critical damping coefficient.

There are several advantages to developing an accurate FEM simulation to evaluate cantilever performance in the PA chamber. With an established framework in place, both traditional rectangular and nontraditional paddle and dual-anchor cantilever designs could be evaluated to optimize the next generation THz PA sensor. Wang et al. [27] used paddle designs out of a silicon nitride film to create airflow sensors. The paddle designs had a narrow anchor width to reduce the spring constant and a large tip surface area to capture the airflow across the sensor. Although a different cantilever design could use a dual-anchor configuration, two narrow anchor widths are spaced apart and connected to a large area to capture pressure waves. For a fixed device layer thickness, paddle and dual-anchor designs may present an advantageous means of modifying the spring constant of the beams by changing the anchor width while maintaining a larger tip surface area to capture the PA pressure.

The analysis of the analytical equations and FEM results highlights the two factors that were considered in selection of a beam design for fabrication: (1) the magnitude of the tip displacement and (2) the cantilever's resonant frequency. If the PA system is modulated at the resonant frequency of the cantilever, a higher resonant frequency would be advantageous to enable a greater number of signal averages over a short duration. A counterargument is to select a reduced resonant frequency design due to Eq. (5), which shows that the anticipated change in chamber pressure per cycle is inversely proportional to the modulation frequency, and a lower resonant frequency means a lower spring constant and therefore greater deflection. From several considered beam designs, the 5 × 2 × 0.01 mm^3 cantilever was selected for fabrication due to the predicted moderate tip deflections and resonant frequency of 550 Hz from the FEM prediction.

http://dx.doi.org/10.5772/62571

3. Device fabrication

The entire process from the initial sensor device concept design through the completed fabrication of the cantilever sensors was performed in this work. The cantilever beam designs were constructed using the device layer of Ultrasil silicon-on-insulator (SOI) (100) wafers. The selection of the SOI device layer and buried oxide (BOX) layer thicknesses were important parameters that affect both the cantilever performance characteristics and manufacturability of the devices. The BOX layer was used as an etch stop material for both the front and backside plasma etches used to create the cantilever devices.

Fabrication began with a thin layer of Ti/Au (20/100 nm) evaporated at the tip of the cantilever, which created the necessary reflective surface for the optical measurement technique employed in the experimental PA tests. A deep reactive ion etch (DRIE) plasma tool was then used to create small 3 to 8 μm gaps that were etched through the device surface of the SIO samples to define the cantilever shape. The majority of the cantilever designs fabricated in this effort were 5 × 2 × 0.01 mm^3. The advantage of using the DRIE is that it is an anisotropic etch process, which results in a near vertical sidewall profile and virtually no mask under cut to create gap widths the same resolution as the photoresist mask pattern. **Figure 3a** shows a schematic layout of the cantilever devices created and **Figure 3b** shows an optical image of a device with a small 3 μm gap achieved with the DRIE process.

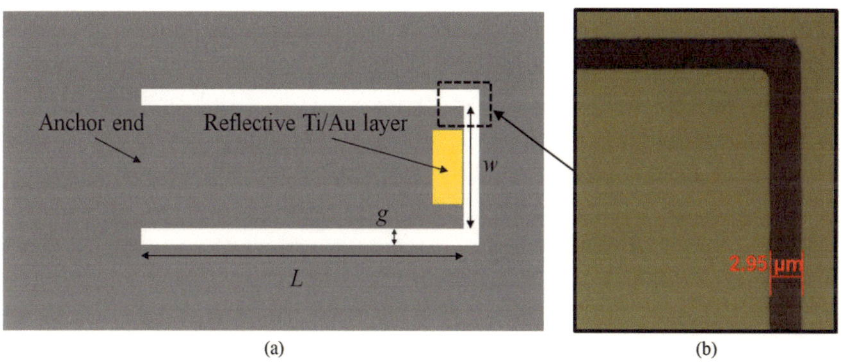

Figure 3. (a) Layout of cantilever design and (b) an optical image of the 3 μm gap etched through the device layer with the DRIE plasma tool.

To the release the cantilever beams, a hole was etched through the backside of the silicon handle wafer in a two-step DRIE and RIE process technique, which was presented in an earlier work [26]. Cantilevers fabricated here had a length/thickness ratio of 500 and a large surface area to be sensitive to the low pressures generated in the THz PA chamber. As a sensor platform, the silicon cantilever performs well, elastically deforming for small cantilever tip deflections. Due to the high length/thickness ratio of the devices, the final stages of the backside etch and cantilever release process proved to be a challenging task. The brittle nature of macroscale silicon was observed when residual stress between the device and BOX layers caused highly localized stress points, which cracked the earlier fabricated cantilever samples. These fabrication challenges were overcome through the use of a photoresist support layer [26] and a thinner 1 μm BOX layer to significantly increase the manufacturability of the high aspect ratio cantilever devices. To complete the fabrication, the BOX under the cantilever was removed with a hydrofluoric acid vapor etch. Completed devices were then tested in the custom THz PA chamber described next in the experimental set-up section.

4. Experimental set-up

The compact custom-fabricated THz PA chamber had overall dimensions of approximately 2 × 2 × 2 in^3 and was constructed out of stainless steel. The test chamber consisted of two segments: a front and back half with the cantilever sensor mounted in between them. The back portion of the chamber contained the absorption cell section, whereas the front half had a small balance volume. The absorption region portion has a cylindrical shape with dimensions of 2 in. long and a diameter of 10 mm. A schematic diagram of the back portion of the PA cell and cantilever position is shown in **Figure 4**.

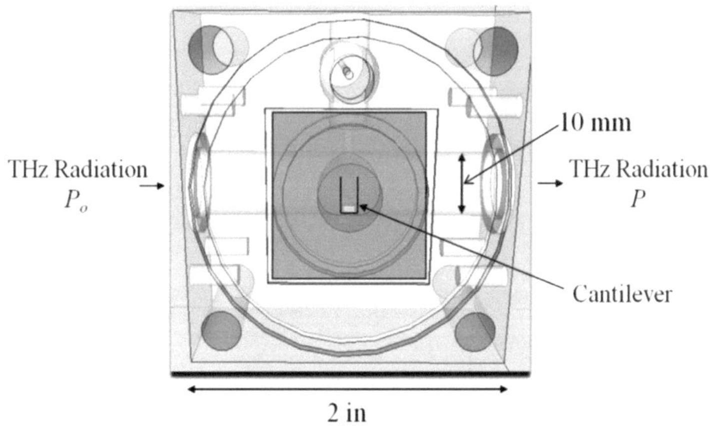

Figure 4. Schematic diagram of PA cell shown with the front portion of the chamber removed; the highlights are chamber dimensions and cantilever position in relation to the absorption cell volume.

To seal the chamber for low-vacuum conditions, Teflon windows were used to enclose the ends of the absorption cell and an antireflective (AR)-coated glass window sealed the balance volume so the optical measurements of the cantilever deflection could be made with a HeNe laser. A Pfeiffer HiCube™ turbo pumping station was used to evacuate the chamber and achieve a low base pressure vacuum level. CH_3CN, also called acetonitrile, was the first gas used to characterize the cantilever sensor. Liquid CH_3CN was exposed to the low-vacuum environment and allowed to vaporize. Through a series of valves, the CH_3CN vapor was allowed into the acoustic cell in a highly controlled manner and the absolute chamber pressure was continuously monitored with a MKS Baratron® capacitance monometer vacuum gauge.

http://dx.doi.org/10.5772/62571

PA data collection was controlled through a series of LabView VI's and the signals were collected with a National Instruments (NI) USB-6221 multifunction data acquisition (DAQ) card. To generate the THz radiation and cause the PA effect, Virginia Diodes, Inc. (VDI) amplifier multiplier chains were used. The signal to the VDI THz radiation diode was provided by an Agilent E8254A PSG-A signal generator. Controlled through the LabView interface, the output of the signal generator was set to a specified THz radiation frequency, which was then amplitude modulated on and off with a 50% duty cycle square wave at the desired modulation frequency. The emitted power by the THz VDI source ranged from 0.6 to 1.4 mW and is frequency dependent. At low chamber pressures, the output power of the THz source was too high and therefore had to be attenuated at low pressures to prevent molecular saturation. On the opposite side of the chamber from the THz source, there is a VDI detector positioned to monitor the transmitted power that exits the chamber and simultaneously measure the absorption spectra with traditional techniques. For this experiment, the amplifier chain used generated radiation over the 0.250 to 0.375 THz frequency range. An advantage to the test set-up is that additional amplifier multiplier chains can be adapted to the system to reach higher frequencies, up to 1.5 THz if required.

The PA chamber and optics for the experimental set-up were mounted on an optical bench. Two optical measurement techniques were employed to examine the PA effect. A diagram of the test set-up is shown in **Figure 5**. A HeNe ($\lambda = 633$ nm) laser beam, guided through a series of mirrors, beam splitter, irises, and focusing lens, was reflected off the tip of the cantilever, back to a photodiode where the laser beam power was measured. An iris beam clipping method or optical beam deflection method, similar to Garcia-Valenzuela and Villatoro [28], was used to collect PA spectral data in the THz test chamber when the reference mirror was blocked.

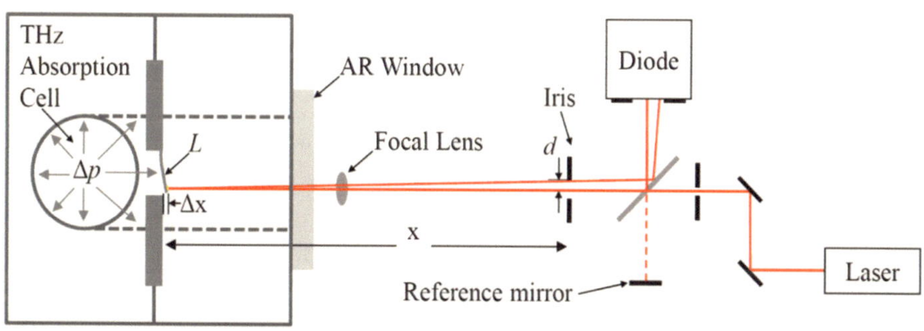

Figure 5. Diagram (not to scale) shows both the PA optical measurement techniques used; beam clipped by the iris when the reference mirror is blocked and a Michelson interferometer displacement measurement when the reference arm was used.

To quantify the sinusoidal changes in power observed at the photodiode, a Michelson interferometer measurement was also incorporated into the experiment. Switching to the interferometer measurement was accomplished by unblocking the reference mirror, modifying the filter settings of the photodiode amplifier, and removing the iris closest to the PA chamber. Accurate cantilever tip deflections measurements were made through the generated constructive and destructive inference signal measured at the diode detector. For large cantilever deflections, the photodiode signal goes through maxima to minima interference patterns as the cantilever deflects a distance of $\lambda/4$.

In both measurement techniques, the photodiode signal was sent to a Stanford Research Systems SR560 preamplifier, set-up in a band-pass filter configuration. The photodiode signal was also sent to a Stanford Research Systems SR530 lock-in amplifier. The amplitude modulated frequency from the Agilent E8254A signal generator is used as the reference signal for the SR530 lock-in amplifier. The two-channel output of the lock-in amplifier reported the magnitude and phase components of the amplified photodiode signal compared to the reference.

For the iris beam clipping method, the measured PA signal at the diode was greatly affected by the iris placement at distance x in front of the cantilever and the positioning of the focal lens. The focal lens was positioned such that the beam formed a tight spot at the tip of the cantilever and beam spot at the detector was smaller than the detector opening. Iris placement in front of the cantilever served two purposes. The first function was to reduce the diameter of the incoming beam before the laser light impinged the cantilever. The second purpose was to act again as an aperture; light reflected off the deflected cantilever was then clipped by the iris before it reached the photodiode. The diameter of the iris closest to the chamber was set to ~0.8 mm. The final spot size of the HeNe laser beam at the photodiode was ~1.1 mm in diameter. The beam displacement d from steady state shown in **Figure 5** is expressed by

$$d = x\tan\left(\sin^{-1}\frac{\Delta x}{L}\right) \tag{13}$$

where x is the distance from the cantilever steady-state position to the iris, Δx is the cantilever deflection distance, and L is the cantilever length. To obtain the optimal signal, the THz source was modulated at the resonant frequency of the cantilever, and the chamber, mounted on a three-axis stage, was adjusted until the photodiode signal produced a symmetric amplitude sine wave. The sinusoidal peak-to-peak voltage signal from the photodiode and the magnitude of the lock-in amplifier signal were used to collect the PA spectral data from the sample gas. The peak-to-peak voltage signal created at the photodiode was due to the linear region of a three-dimensional Gaussian beam profile caused by the clipping iris when the center of the beam was shifted by a total distance of $2d$. The PA signal and interferometer measurements are discussed further in the Results section.

5. Results

The experimental results are discussed in two distinct topic areas: cantilever behavior and PA spectral data. The first important aspect to discuss is the cantilever sensor dynamic behavior in response to the PA effect in the test chamber. Relevant information about the behavior of the cantilever sensor includes determining the resonant frequency, quality factor, time constants, and PA signal vs. cantilever deflection for the system over the range of tested pressures. Based on the identified cantilever sensor traits, data collection rates, techniques, and PA spectral collection results are discussed.

5.1. Cantilever sensor

To investigate the cantilever and PA chamber performance, CH_3CN was selected because it has documented absorption lines over a large spectrum (0.018–1.8065 THz), making it an ideal gas, as the radiation source can span a large portion of that spectrum. Also, CH_3CN has both strong and weak absorption line strengths within that frequency range, which would demonstrate the large dynamic range capability of the cantilever PA system. Initial PA data were taken using the absorption line frequency just over 312.633 GHz, identified as Freq 1 from here forward. Then, to acquire the optimal PA signal from the cantilever, a modulation frequency scan was performed using the THz Freq 1 absorption line. Modulation frequency scans from 1 to 700 Hz found that the maximum PA signal was achieved when the system was modulated at the resonant frequency ($\omega \cong \omega_o$) of the cantilever. **Figure 6** shows the PA signal from the lock-in amplifier as a function of modulation frequency over the 3 to 400 mTorr pressure ranges.

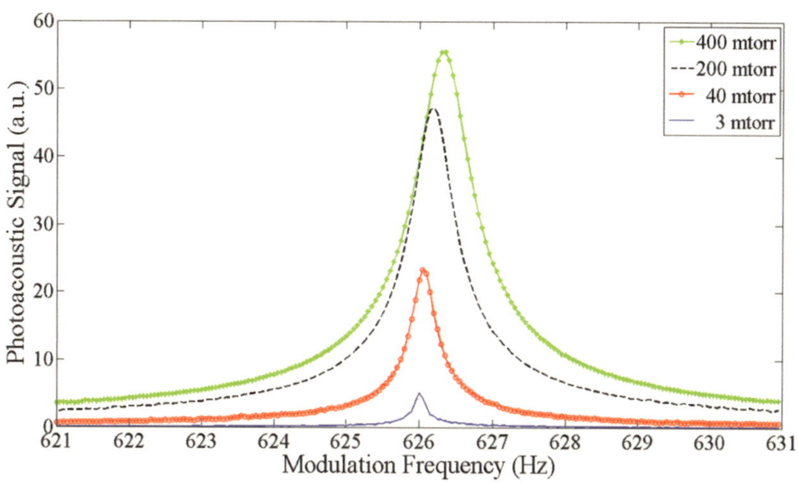

Figure 6. PA response of the cantilever as a function of modulation frequency for four different pressures tested revealed that the maximum PA signal was achieved at slightly higher modulation frequencies with increased chamber pressure.

The required modulation frequency to achieve the maximum PA signal increased nonlinearly with increasing chamber pressure, approximately 8.8×10^{-4} Hz $mTorr^{-1}$ overall. The slight observed frequency shift was attributed to an increase in the effective spring constant of the system due to the gas interactions with the cantilever, which increased as a function of chamber pressure [20]. More importantly, from the modulation frequency scans, the quality factor of the cantilever in the low-vacuum conditions can be assessed. The quality factor of a cantilever can be expressed as

$$Q = \frac{\omega_o}{\Delta\omega} = \frac{f_o}{\Delta f} \tag{14}$$

where f_o is the resonant frequency of the cantilever and Δf is the full-width half-maximum of the PA signal. Due to the low damping conditions of the rarified gas in the PA chamber, the cantilever had a sharp resonance response to the modulation frequency scan and small Δf values of 0.15 to 1 Hz were measured. At the low vacuum pressures tested, the 5 × 2 × 0.01 mm^3 cantilever design had high-quality factors that are shown in **Figure 7**.

Using the optimal modulation frequency of the cantilever found at each pressure, the dynamics of cantilever excitation and relaxation was investigated. **Figure 8** shows a sample of PA data collected from the system at a chamber pressure of 3 mTorr using THz Freq 1. The THz excitation on time was 8 s and it can be seen in the graph that the lock-in PA signal was just approaching a steady state before the radiation was turned off. As the chamber pressure increased, the PA signal response time improved. In **Figure 9**, PA data were collected at a chamber pressure of 80 mTorr. The higher chamber pressures showed a much quicker response to reaching steady-state conditions and generated a stronger PA signal, and the output of the lock-in signal reached a steady-state value in approximately 4 s.

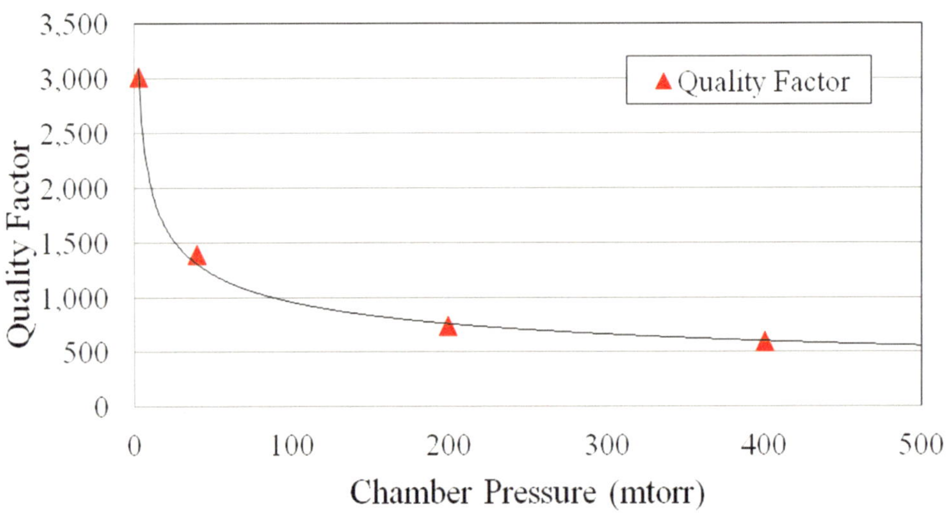

Figure 7. Quality factor of the 5 × 2 × 0.01 mm^3 cantilever sensor was high due to the low-pressure intrinsic/molecular damped environments in the PA chamber.

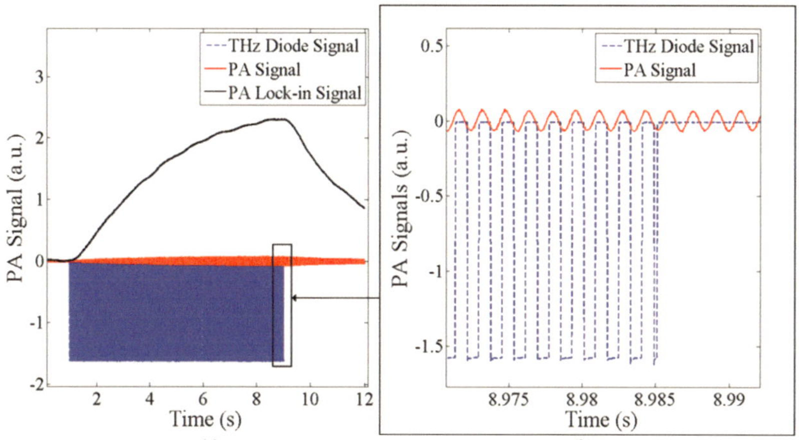

Figure 8. PA data recorded at 3 mTorr in plot (a) shows the slow response of the lock-in amplifier and raw PA signals at the low chamber pressure. Plot (b) is a zoomed-in view of the raw PA signal from the diode and modulated THz signal as the radiation is turned off.

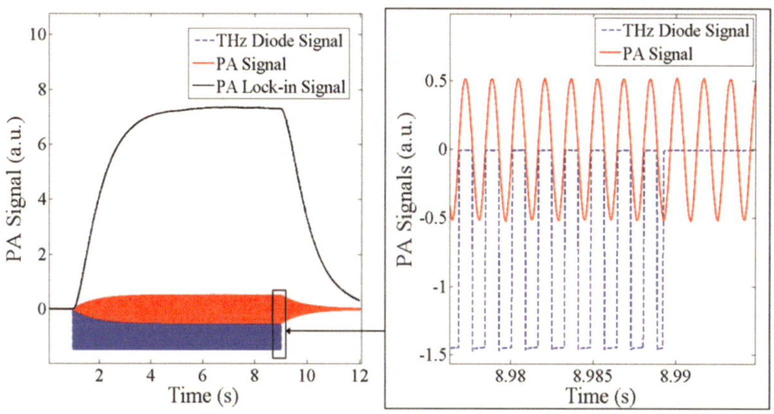

Figure 9. PA data in plot (a), taken at 80 mTorr, illustrates the quicker response time of the lock-in PA signal and raw PA signal from the THz excitation radiation at higher chamber pressure. Plot (b) is a zoomed-in view of the raw PA signal from the diode and modulated THz signal.

For a range of pressures, the cantilever was brought up at an excited state and then the exponential decay of the PA signal was recorded. An exponential curve fit was then performed on the PA data to extract the time constants, τ. The resulting time constant versus pressure plot is shown in **Figure 10**. Time constants dramatically increase as the chamber pressure gets lower and moves from the molecular to intrinsic pressure regime. For the investigated pressures, a curve fit was performed using a power function. An equation for the time constant as a function of pressure was found to be

$$\tau(p_c) = 4.553 p_c^{-0.32556} \tag{15}$$

where the chamber pressure p_c is given in mTorr. The resultant function shown in Eq. (15) had a strong correlation to the experimental measurements and had an R^2 value of 0.998. At a chamber pressure of 3 mTorr and a lock-in time constant of 0.3 s, it was found to take approximately 8 s for the PA signal from the lock-in amplifier to reach a steady-state value. This long required excitation time at lower chamber pressures can be expected due to the extremely low pressure change generated per modulation pulse.

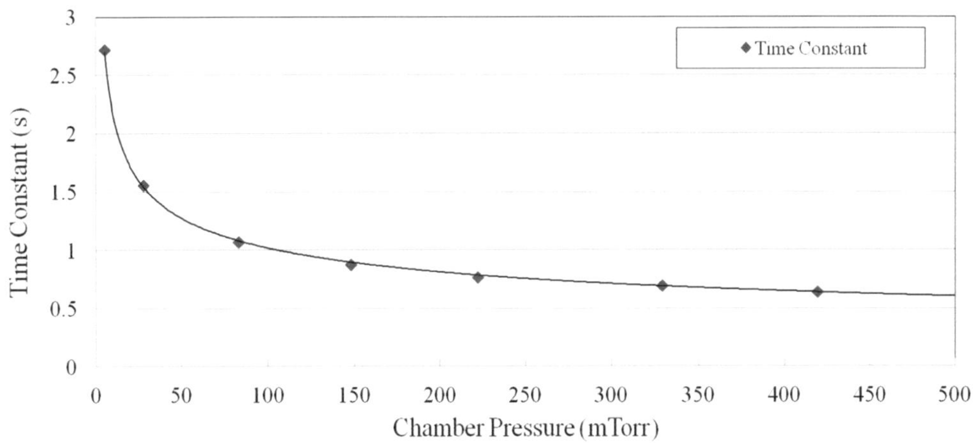

Figure 10. Time constants found through the exponential curve fit to the decay of PA signals over the range of pressures shown.

To quantify the PA signal generated by the diode voltage measurements, cantilever displacement measurements with a Michelson interferometer configuration were performed. Peak-to-peak PA signal and cantilever tip displacement measurements shown in **Figure 11** were taken across a small range of pressures. The left side of the graph corresponds to the diamond-shaped markers, which are the peak-to-peak PA signal data points. Circular data points on the graph with error bars correspond to the axis on the right, which shows the amplitude of cantilever deflection from center position given in microns. When the chamber was pumped down to a high vacuum condition, no THz PA signal could be measured. It can be seen in the graph that, as the pressure increased with the addition of CH_3CN, the measured PA signal and cantilever deflection continued to increase with pressure. Above ~7 μm deflection, the PA signal increased only slightly due to the positioning of the laser beam through the iris. This strong correlation between the PA signal and deflection measurements provides good justification for using the iris-clipped PA method for the range of deflections encountered. An additional advantage of the of the iris-clipped PA method is the modest required sampling rate; four channels of data can be digitized at 60,000 samples per second allowing for multiple channels of data to be collected simultaneously.

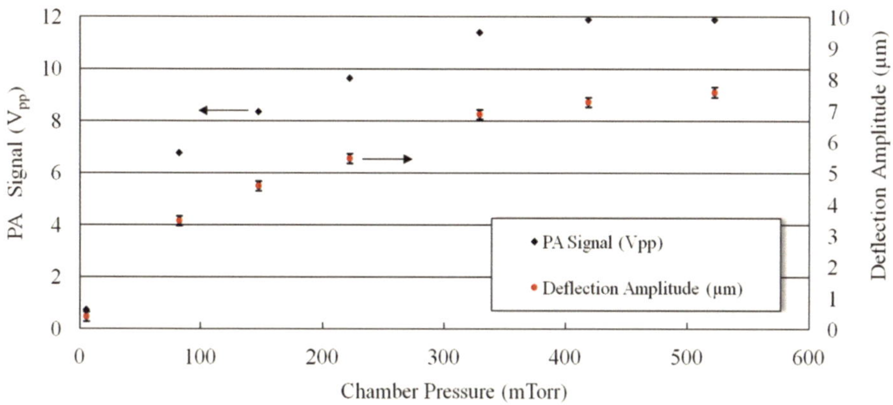

Figure 11. Graph of measurements taken on 5 × 2 × 0.01 mm^3 cantilever at different chamber pressures show PA peak-to-peak signals represented by a diamond shape corresponding to the scale on left, whereas the interferometrically measured cantilever amplitude deflections, shown with a circle and error bars, correspond to the deflection scale on the right.

The frequency spectrum of the PA system noise and an active PA signal were also investigated. In **Figure 12**, a 1 s duration of data was evaluated using Fourier analysis of the PA signal and noise. The inset in **Figure 12** shows the single-sided amplitude of the FFT when the THz source was off. With the THz source operated at Freq 1 at a modulation frequency of 626.02 Hz, the PA signal was more than 1000 times greater than the noise. The PA system had a low amount noise in the measurement signal. Two portions of the spectrum that stood out slightly were the noise signals occurring below 200 Hz and region around the resonant frequency of the cantilever. The signal around the resonant frequency of the cantilever beam was attributed to the pickup of vibrations in the experimental set-up due to noise and vibrations in the room.

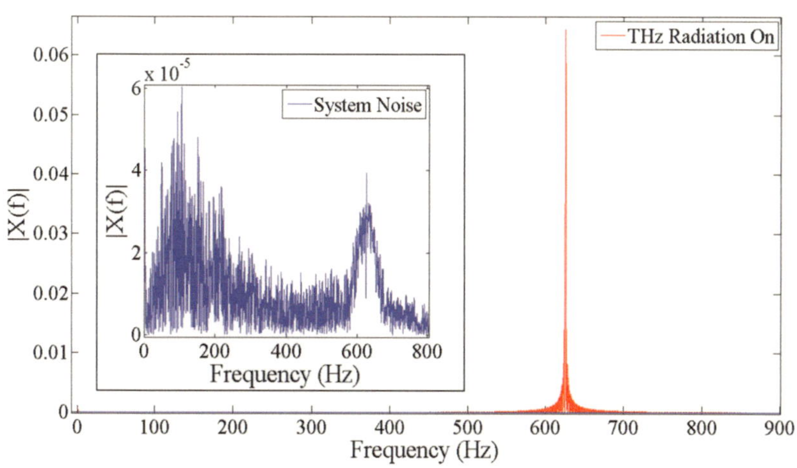

Figure 12. Fourier analysis of a 1 s PA signal, collected at 200 mTorr taken at Freq 1, showed low noise in the system with no single dominant noise frequency in the iris-clipped PA measurement method.

5.2. PA spectral analysis

PA spectral data collection was also performed across a range of pressures. As discussed earlier in the cantilever sensor analysis, chamber pressure greatly affected PA signal strength and response time. Therefore, two data collection techniques were investigated; fast scan methods to coarsely cover a broader-frequency spectrum and a slow scan method to achieve a stronger and more accurate PA signal. Spectral data collection measurements were performed in two steps: excitation and signal averaging segments. The amplitude modulated THz radiation at the specified THz frequency was turned on for a period of time, which we defined as the excitation time. This excitation time or period allowed the cantilever to attain an excited level of amplitude displacement from the generated PA pressure pulses. After the excitation time, the PA signal was averaged over the second specified interval, referred to as the signal averaging period. The PA signal measured during the signal averaging period was recorded as the PA signal for that specific THz frequency. The THz frequency was then increased by a small amount and the measurement sequence was repeated.

To show the effects of excitation time and resulting PA signal, different excitation times were performed. The signal averaging period was held constant for each of the trials and was set to a 0.1 s duration. **Figure 13** shows the results of three excitation times taken at a chamber pressure of 59 mTorr with a frequency step size of 0.05 MHz. The 2 s excitation PA spectral collection took 9.7 min, the 0.5 s excitation set took 2.8 min, and the 0.1 s excitation data set took 0.9 min to collect. The PA response shows a pronounced frequency shift at the rapid 0.1 s excitation time and a marked reduction in PA signal strength. This observed frequency shift highlights three regions of particular interest in the PA signal. First is the frequency location of the maximum PA signal, and it should ideally occur at line center frequency where the

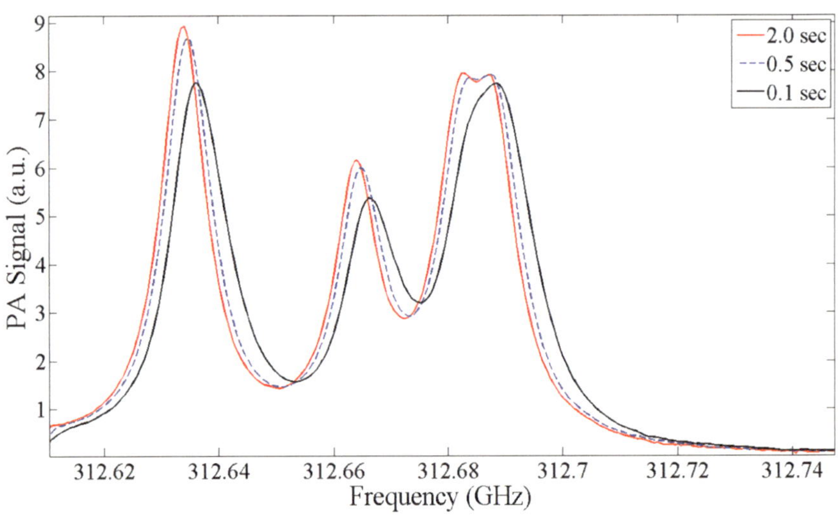

Figure 13. Spectral PA signals taken at 59 mTorr with 2, 0.5, and 0.1 s excitation times using a 0.05 MHz step size, moving from low to high frequency, the PA peak response shifted to higher frequencies and reduced amplitudes as the excitation time was decreased.

maximum absorption occurs. The other two important regions are the rising and the falling edge of the absorption curve about the maximum PA signal, which are due to the line broadening mechanisms mentioned earlier.

A few drawbacks exist to low-pressure (2–60 mTorr) spectral PA data collection. On the rising edge of the PA signal, PA excitation to full amplitude deflection under low-pressure conditions can take more than 4 to 12 s due to the low excitation pressure generated in the chamber and the high-quality factor of the cantilever resonator. Likewise, on the falling edge of the PA signal, once the cantilever reaches an excited steady state, the small damping coefficient at the lower chamber pressures and continued excitation at a lower absorption strength cause the amplitude of the cantilever deflection to reduce slowly, which is manifested in a small frequency shift on the downward slope of the PA signal. Radiation excitation frequency and the corresponding PA response are the crucial parameters that would be used for chemical identification applications, so they must be accurately identified.

That being said, the fast scan technique tested above is extremely useful and could be used to quickly evaluate small spectral regions for absorption lines if an unknown chemical were evaluated with the system. Then, a slower, more accurate scan could be performed to identify the line center absorption frequency and the relative PA amplitudes for each absorption line. To achieve an accurate line center absorption frequency during a spectral scan, the excitation time had to be increased as the chamber pressure decreased. At the lower 2 to 5 mTorr pressures tested, spectral line broadening appeared to be dominated by Doppler broadening effects and created the narrowest spectral line widths. As line broadening increased with pressure, the frequency step size could be increased and the excitation time could be decreased. In **Figure 14**, a small portion of spectra was recorded for three low-pressure cases. Each collection took 10 to 12 h due the small 0.05 MHz step size used and the 12

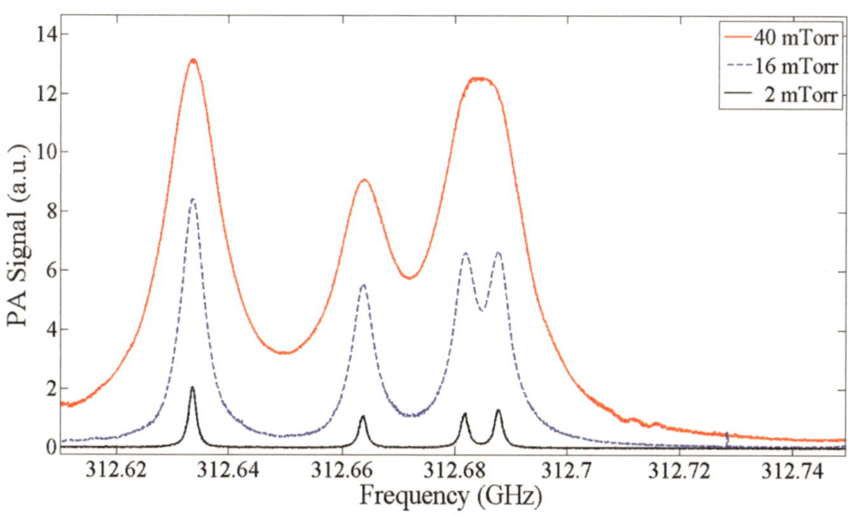

Figure 14. PA spectral data of CH_3CN collected at three pressures used a 0.05 MHz step size and recorded the 0.5 s average PA signal for each frequency step.

to 15 s excitation time before the subsequent 0.5 s average PA signal was recorded. The PA spectra presented in **Figure 14** is of interest for two reasons. This spectral region contains strong absorption lines and it also contains two peaks in close proximity, separated by only 6 MHz. The data in **Figure 14** also show the effects of pressure broadening in the PA signal as the two closely spaced absorption lines merge into one PA peak at higher chamber pressures.

During the long data collection runs, the PA measurement scheme employed for the experiment using the iris-clipped laser beam diode signal and lock-in amplifier signals proved to be extremely stable. The PA chamber maintained low vacuum levels well. Over the course of the long data collections, the chamber pressure increased slowly and had an average system leak rate of 0.5 to 1 mTorr h^{-1}.

To demonstrate the broader spectral performance of the THz PA system, a 0.72 GHz frequency scan was performed spanning 312.07 to 312.79 GHz. This low-pressure PA signal taken at 13 mTorr successfully captured the 13 rotational absorption lines in the CH_3CN spectral region scanned. The data collection method used 0.2 MHz step size and 12 s excitation, and then the 0.5 s average signal from the lock-in amplifier was recorded. This much larger PA scan of CH_3CN shown in **Figure 15** took 12.5 h to collect, and it also compares the simultaneously recorded THz diode signal to the modeled absorption coefficient profile. The scaled simulated spectra closely matched the measured PA signature of the gas with a small deviation at higher absorption strengths.

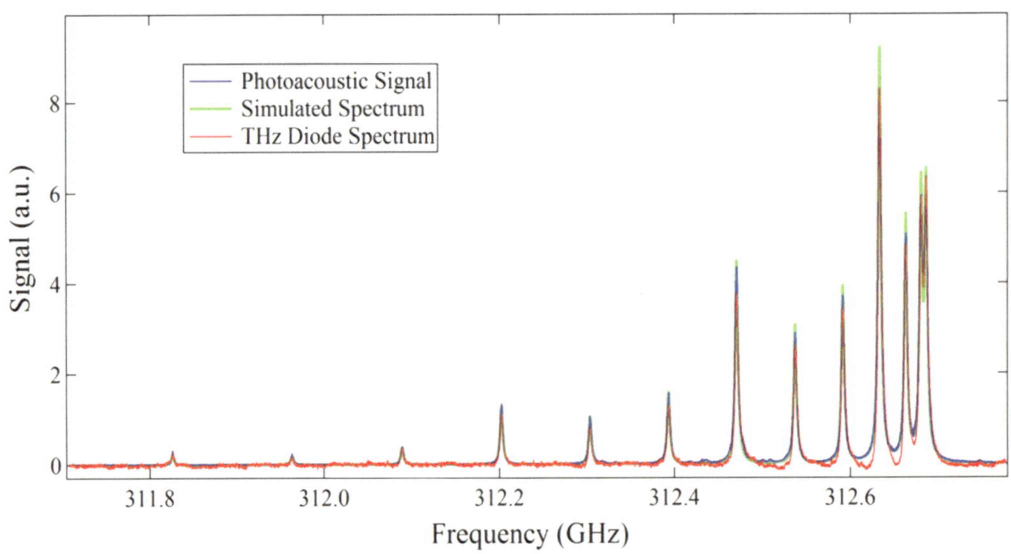

Figure 15. PA data and the simultaneously recorded THz diode signal are compared to the simulated absorption spectra of CH_3CN recorded at 13 mTorr.

With excellent PA spectral data collected at multiple low pressures, a further analysis of the system performance parameters was carried out. SNR, sensitivity, and NNEA coefficient of

http://dx.doi.org/10.5772/62571

the PA sensor were important figures of merit for system performance. Data collected at 13 mTorr with the THz source power attenuated down to ~0.1 mW had a peak PA signal of 7.084, measured at 312.6336 GHz. The RMS noise floor for the collected data set was determined to be 0.0058, which was located between two weak absorption peeks at 311.925 GHz. Using the peak and minimum PA signals in the spectral data collection, the SNR = 1221. Using Eqs. (2) and (3), a sensitivity of 1.97×10^{-5} cm^{-1} and an NNEA of 1.39×10^{-9} cm^{-1} W $Hz^{-½}$ were determined using a 0.5 s signal averaging time. The sensitivity and NNEA achieved in this effort were excellent and compare well to previously reported cantilever PA trace gas detection systems while operating at significantly lower chamber pressures.

6. Conclusions

In this work, a sensitive MEMS cantilever design was successfully modeled, fabricated, and tested in a custom-designed low-vacuum THz PA chemical sensing and molecular spectroscopy system. Cantilever design parameters of length, width, and thickness of the cantilever structure were used to select a cantilever design that was sensitive to the generated PA pressures in the chamber. With this novel system, the first ever cantilever-based THz PA spectrum was collected and analyzed at low chamber pressures in the 2 to 60 mTorr range. The pressure regimes tested here were several orders of magnitude lower than the trace gas detection cantilever PA systems discussed earlier [11–18]. Significant PA signals were achieved at chamber pressures as low as 2 mTorr. For THz spectroscopy applications, the sensor size and chamber pressures tested were also an order of magnitude smaller than the membrane sensor system used by Krupnov and Burenin [9]. By collecting PA spectra in the 2 mTorr pressure regime, it allowed for a primarily Doppler limited line broadening of the CH_3CN, making closely spaced absorption lines highly resolvable.

This research effort achieved an NNEA of 1.39×10^{-9} cm^{-1} W $Hz^{-½}$ using a compact THz PA spectroscopy and chemical sensing system. Due to its compact size, the system could be used as a portable chemical sensing and spectroscopy platform. This would be a great advantage in comparison to a large traditional absorption cell for spectroscopy applications and may lead one day to a handheld THz chemical sensor or MEMS detector arrays for THz imaging applications.

Acknowledgements

The authors thank the Air Force Office of Scientific Research (AFOSR) for funding this effort (F4FGA04013J001) and the Air Force Research Laboratory (AFRL) Sensors Directorate for their assistance. The authors also thank the AFIT clean-room staff, Mr. Richard Johnston, and Mr. Thomas Stevenson.

Author details

Nathan Glauvitz[1], Ronald A. Coutu[1*], Ivan R. Medvedev[2] and Douglas T. Petkie[2]

*Address all correspondence to: Ronald.Coutu@AFIT.edu

1 Air Force Institute of Technology, Hobson Way, Wright-Patterson AFB, USA

2 Wright State University, Dayton, USA

The views expressed in this paper are those of the authors and do not reflect the official policy or position of the United States Air Force, Department of Defense, or the U.S. Government.

References

[1] West G. A., Barrett J. J., Siebert D. R., and Reddy K. V. Photoacoustic spectroscopy. *Rev. Sci. Instrum. 54*, pp. 797–817, 1983.

[2] Bell A. G. *Upon the Production of Sound by Radiant Energy.* Washington, DC: Gibson Brothers Printers, 1881.

[3] Miklos A., Hess P., and Bozoki Z. Application of acoustic resonators in photoacoustic trace gas analysis and metrology. *Rev. Sci. Instrum. 72*, pp. 1937–1955, 2001.

[4] Brugel W. *An Introduction to Infrared Spectroscopy.* New York: John Wiley & Sons, Inc., 1962.

[5] Kosterev A. A., Bakhirkin Y. A., Curl R. F., and Tittel F. K. Quartz-enhanced photoacoustic spectroscopy. *Opt. Lett. 27*, pp. 1902–1904, 2002.

[6] Kosterev A. A., Tittel F. K., Serebryakov D. V., Malinovsky A. L., and Morozov I. V. Applications of quartz tuning forks in spectroscopic gas sensing. *Rev. Sci. Instrum. 76*, pp. 043105, 2005.

[7] Liu K., Li J., Wang L., Tan T., Zhang W., Gao X., Chen W., and Tittel F. K. Trace gas sensor based on quartz tuning fork enhanced laser photoacoustic spectroscopy. *Appl. Phys. B 94*, pp. 527–533, 2009.

[8] Borri S., Patimisco P., Sampaolo A., Beere H. E., Ritchie D. A., Vitiello M. S., Scamarcio G., and Spagnolo V. Terahertz quartz enhanced photo-acoustic sensor. *Appl. Phys. Lett. 103*, pp. 021105, 2013.

[9] Krupnov A. F. and Burenin A. V. New methods in submillimeter microwave spectroscopy. In *Molecular Spectroscopy: Modern Research*, Rao K. N., ed. New York: Academic Press, Inc., 1976, pp. 93–126.

[10] Firebaugh S. L., Jensen K. F., and Schmidt M. A. Miniaturization and integration of photoacoustic detection with a microfabricated chemical reactor system. *J. Microelectromech. Syst. 10*, pp. 232–237, 2001.

[11] Ledermann N., Muralt P., Baborowski J., Forster M., and Pellaux J. Piezoelectric pb(zrx, ti1-x)O_3 thin film cantilever and bridge acoustic sensors for miniaturized photoacoustic gas detectors. *J. Micromech. Microeng. 14*, pp. 1650–1658, 2004.

[12] Kuusela T., Peura J., Matveev B. A., Remennyy M. A., and Stus N. M. Photoacoustic gas detection using a cantilever microphone and III-V mid-IR LEDs. *Vibrat. Spectrosc. 51*, pp. 289–293, 2009.

[13] Kuusela T. and Kauppinen J. Photoacoustic gas analysis using interferometric cantilever microphone. *Appl. Spectrosc. Rev. 42*, pp. 443–474, 2007.

[14] McNaghten E., Grant K., Parkes A., and Martin P. Simultaneous detection of trace gases using multiplexed tunable diode lasers and a photoacoustic cell containing a cantilever microphone. *Appl. Phys. B Lasers Opt. 107*, pp. 861–871, 2012.

[15] Adamson B. D., Sader J. E., and Bieske E. J. Photoacoustic detection of gases using microcantilevers. *J. Appl. Phys. 106*, pp. 114510, 2009.

[16] Fonsen J., Koskinen V., Roth K., and Kauppinen J. Dual cantilever enhanced photoacoustic detector with pulsed broadband IR-source. *Vibrat. Spectrosc. 50*, pp. 214–217, 2009.

[17] Peltola J., Vainio M., Hieta T., Uotila J., Sinisalo S., Metsälä M., Siltanen M., and Halonen L. High sensitivity trace gas detection by cantilever-enhanced photoacoustic spectroscopy using a mid-infrared continuous-wave optical parametric oscillator. *Opt. Express 21(8)*, pp. 10240–10250, 2013.

[18] Kauppinen J., Wilcken K., Kauppinen I., and Koskinen V. High sensitivity in gas analysis with photoacoustic detection. *Microchem. J. 76*, pp. 151–159, 2004.

[19] Sievilä P., Chekurov N., Raittila J., and Tittonen I. Sensitivity-improved silicon cantilever microphone for acousto-optical detection. *Sens. Actuators A Phys. 190*, pp. 90–95, 2013.

[20] Koskinen V., Fonsen J., Roth K., and Kauppinen J. Progress in cantilever enhanced photoacoustic spectroscopy. *Vibrat. Spectrosc. 48*, pp. 16–21, 2008.

[21] Ariyoshi S., Otani C., Dobroiu A., Matsuo H., Sato H., Taino T., Kawase K., and Shimizu H. M., Superconducting detector array for terahertz imaging applications. *Jpn. J. Appl. Phys. 45*, pp. L1004–L1006, 2006.

[22] Tao H., Kadlec E. A., Strikwerda A. C., Fan K., Padilla W. J., Averitt R. D., Shaner E. A., and Zhang X. Microwave and terahertz wave sensing with metamaterials. *Opt. Express 19*, pp. 21620–21626, 2011.

[23] Chen S., Choe Y., Baumgartel L., Lin A., and Kim E. S. Edge-released, piezoelectric MEMS acoustic transducers in array configuration. *J. Micromech. Microeng. 22*, pp. 025005, 2012.

[24] Blom F. R., Bouwstra S., Elwenspoek M., and Fluitman J. H. J. Dependence of the quality factor of micromachined silicon beam resonators on pressure and geometry. *J. Vac. Sci. Technol. B 10*, pp. 19–26, 1992.

[25] Uotila J., Koskinen V., and Kauppinen J. Selective differential photoacoustic method for trace gas analysis. *Vibrat. Spectrosc. 38*, pp. 3–9, 2005.

[26] Glauvitz N. E., Coutu R. A., Kistler M., Medvedev I. R., and Petkie D. T. MEMS cantilever sensor for photoacoustic detection of terahertz radiation. In *Proceeding of SEM 2013 Annual Conference & Exposition on Experimental and Applied Mechanics*, Lombard, IL, 2013.

[27] Wang Y., Lee C., and Chiang C. A MEMS-based air flow sensor with a free-standing micro-cantilever structure. *Sensors 7*, pp. 2389–2401, 2007.

[28] Garcia-Valenzuela A. and Villatoro J. Noise in optical measurements of cantilever deflections. *J. Appl. Phys. 84*, pp. 58–63, 1998.

Chapter 3

Photonic Crystal Chemical/Biochemical Sensors

Saeed Olyaee, Hamideh Mohsenirad and
Ahmad Mohebzadeh-Bahabady

Additional information is available at the end of the chapter

http://dx.doi.org/ 10.5772/63288

Abstract

In this chapter, the definitions of the photonic crystals (PhCs) are presented and the photonic crystal-based sensors are described. The structures of some photonic crystal biosensors that can detect chemical or biochemical molecules are also investigated. Sensing mechanism in the most photonic crystal sensors is based on the refractive index (RI) change mechanism. By binding the chemical or biochemical molecules to active sensing surface, the refractive index will be changed. So, the resonant wavelength or the intensity of the transmission spectrum is changed. This process can be used as a way to measure the concentration of the molecules. To simulate the optical wave behavior in the structure and evaluating the ability of biosensing, the two-dimensional finite-difference time-domain (2D-FDTD) and plane-wave expansion (PWE) methods are used. The sensors that are presented in this chapter are mostly based on photonic crystal resonators. Some structures based on micro-/nano-resonators, structures based on LX resonators, and structures based on ring resonator are investigated. Important parameters on sensing applications for these structures are also calculated.

This chapter is organized as follows: an introduction to photonic crystal structures is presented in Section 1. Then, sensing mechanism in chemical sensors based on photonic crystals is described in Section 2. Quality factor, full width at half maximum (FWHM), free spectral range (FSR), detection limit (DL), and sensitivity are the most important parameters in sensing applications which are introduced in Section 3. Finally, the last section describes some structures of photonic crystal chemical/biochemical sensors.

Keywords: photonic crystal, nanostructures, chemical sensor, biochemical sensor, micro-/nano-resonator

1. Introduction

In recent years, designs of optical devices based on periodic structures have been widely developed. One of the periodic structures that can control and guide the photons is photonic crystal (PhC). Photonic crystals are the periodic dielectric structures that can be categorized into three main groups as one-dimensional (1D), two-dimensional (2D), and three-dimensional (3D) structures [1].

A good example of 1D PhCs is the Bragg grating which the periodic dielectrics of structures occur in one direction, while in two other directions structure is uniform. In 2D PhC structure, the periodicity of the dielectrics occurs in two directions. An example of the 2D PhC is the periodically arranged system of air holes in the silicon substrate or a lattice of dielectric rods in air. If three dimensions in the photonic crystal structure have permittivity modulation, the structure will be called 3D PhC. 1D, 2D, and 3D PhC structures are shown in **Figure 1**.

Photonic band gap (PBG) is the important property of the photonic crystals. PBG mentions to forbidden frequency or wavelength range in PhC structure. However, a little change in the periodic structure, which is called as defect, can create a new frequency range in the band gap [1]. In the lattice of air holes in silicon, defects can be formed by reducing or increasing the radius of one or more air holes; even defects can be shaped by filling holes.

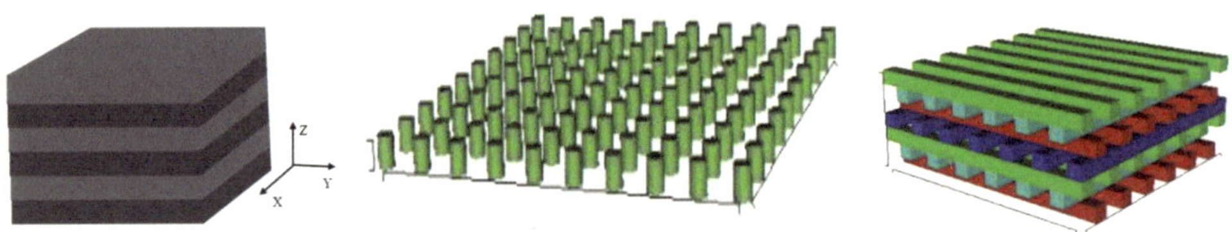

Figure 1. Photonic crystal structures (a) 1D, (b) 2D, and (c) 3D [1].

The photonic crystals have been presented by various structures and have been used in various applications. These structures such as fibers, waveguides, cavities, micro- and nano-resonators, and ring resonators and also their applications including photonic crystal multiplexers, photonic crystal filters, photonic crystal lasers, photonic crystal sensors, etc., have been reported.

Photonic crystals can be fabricated by complementary metal–oxide–semiconductor (CMOS) techniques and because of the unique capability to control light in small dimensions are excellent platforms for ultra-sensitive sensors to small changes in the refractive index. These advantages and other benefits caused photonic crystal sensors as a strong sensing platform [2]. Photonic crystal sensors have been designed to detect different parameters such as gas, pressure, displacement, chemical, and biochemical molecules [3–8].

In photonic crystal chemical and biochemical sensors, the analysis is mainly done by two methods: two-dimensional finite-difference time-domain (2D-FDTD) and plane-wave expansion (PWE) approaches. The propagation of electromagnetism wave can be simulated by the 2D-FDTD method and the PBG can be calculated with the PWE approach.

In this section, an overview of the photonic crystal chemical and biochemical sensors is discussed.

2. Sensing mechanism in the photonic crystal chemical sensor

Two detection protocols can be used in optical chemical and biochemical sensors: fluorescent-based detection and label-free sensing [2]. In fluorescent-based detection, the biological molecules are altered. The biological molecules are labeled by fluorescence, and then, the intensity of the sensor output is measured. The measured intensity is used to represent the presence of biological molecules. But, in label-free sensing, the target biological molecules are not labeled and are detected in their natural forms.

The label-free based chemical/biochemical sensors were reported to variety of methods such as surface plasmon resonance, interferometer, waveguide, fiber, ring resonator, and photonic crystal [2]. Among these methods, the method based on the photonic crystals has been more attention.

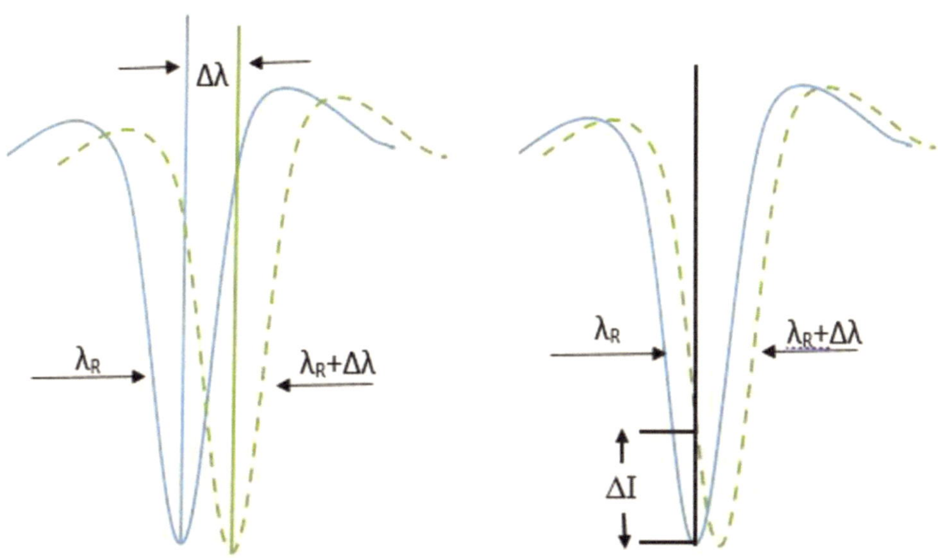

Figure 2. (a) Resonant wavelength shift pattern and (b) intensity variation pattern.

Sensing mechanism in the most photonic crystal sensors is similar in the use of refractive index (RI) change mechanism. The RI change does not depend on the total sample mass but depends on the surface density. When nano-measurement is involved, this characteristic will be very important.

In RI change mechanism, when the biological molecules inject to the optical sensor, the biological molecules bind to active sensing surface. As a result, the refractive index of the active sensing surface will be changed. So, some of the characteristics of the transmission spectrum will be changed. This process can be used as a way to measure the concentration of the biological molecules [9].

In the first category, to identify the presence of biological molecules, the amount of changes in the resonance wavelength is calculated (see **Figure 2(a)**). This method is based on the resonant wavelength shift (RWS) scheme. This scheme is used to measure a wider range that is determined by the free spectral range (FSR). To improve the accuracy of sensor based on RWS-scheme, achieving narrow full width at half maximum (FWHM) is very important. Using this method, the tiny shift in the effective refractive index cannot be measured. For this measurement, high sensitivity sensor is required.

In the second category, to identify the presence of biological molecules, the amount of the intensity change at the resonant frequency is measured. This method is based on the intensity variation (IV) scheme (see **Figure 2(b)**). In this case, a broader FWHM will be desirable and only a small range of wavelength shift can be calculated.

3. Important parameters in sensing applications

Compact optical devices with high quality can be made by precise control in geometric parameters, which some of them have inconsistency. Thus, in biosensing applications, they should be selected so that the optimum state is achieved. Some of the most important parameters of the sensor are as follows.

3.1. Quality factor

Quality factor is the ratio of resonant wavelength (λ_{res}) to the full width at half maximum (FWHM) of drop waveguide output ($\Delta\lambda$). Quality factor can be expressed as $\lambda_{res}/\Delta\lambda$ or $\omega/\Delta\omega_{FWHM}$, where ω is the resonant frequency and $\Delta\omega$ is the full width at half maximum.

3.2. Full width at half maximum (FWHM)

Full width at half maximum is an important parameter to design the sensor, because it affects the quality factor and detection limit (DL). As FWHM be more narrowed, the quality factor and detection limit will be better. With increment of loss, FWHM is increased.

3.3. Free spectral range (FSR)

Free spectral range should be sufficiently wide that adjacent resonant peaks do not interfere in operation of the working resonant peak. FSR is directly proportional to the square of the wavelength. Thus, in the shorter wavelengths, FSR will be smaller. Since the wavelength has a more impact on the FSR, design of sensor in longer wavelengths is the easiest way for optimization of the FSR. In sensors that the presence of the analyte is determined, FSR is not very important parameter. Sensors that use the change density method for the sensing have small FSR. But the sensors that use resonant wavelength shift method with large scale need to wider FSR.

3.4. Detection limit (DL)

Detection limit is the minimal physical change that can be detected. It can be calculated by DL = FWHM/*S*, where *S* is the sensitivity. The detection limit in homogenous sensing can be defined as the smallest changes in the refractive index unit which sensor can detect. Detection limit is minimal increasing the thickness of add-layer that is formed by binding biomaterial to sensing surface.

3.5. Sensitivity

Sensitivity is the wavelength shift per weight of biomolecule in femto-gram unit or, according to another definition, wavelength shift per refractive index unit.

4. Photonic crystal chemical/biochemical sensors

Most of the photonic crystal biosensors are able to detect chemical and biochemical molecules. Biosensors can be designed according to different photonic crystal structures such as photonic crystal fibers [10, 11], photonic crystal-based waveguides [12, 13], photonic crystal-based resonators [14–20], and ring resonators [21–24]. Now, research on the photonic crystal resonator-based sensors has more attention. In the following, some of the biosensors or biochemical sensors have been presented.

4.1. Biosensor or biochemical sensors based on photonic crystal resonators

The biosensors or biochemical sensors usually have a resonator and one or two waveguides. The resonator plays the role of sensing hole. In the resonator, since light is strongly limited, quality factor is improved.

4.1.1. Photonic crystal biosensor with wide measurement range

Olyaee and Najafgholinezhad were introduced a biosensor based on nano-cavity for identification of chemicals and biomaterial molecules such as DNA and protein [14]. The basic structure of the biosensor is illustrated in **Figure 3(a)**. The structure consists of a nano-cavity and two end waveguides. The nano-cavity is obtained by reducing radius of one air hole in the middle of the photonic crystal structure and the two waveguides are formed by eliminating a row of air holes.

In the structure design, many parameters such as the size of nano-cavity hole, the coupling distance, and the length of the input and output waveguides have been calculated. To select the appropriate value for the size of the nano-cavity, the transmission spectra and the quality factor have been studied with for different sizes of defects. As shown in **Figure 3(b)** and **(c)**, for different radii of the defect, the quality factor has a maximum value for the radius equal to 80 nm. In this study, in order to have the high output, $R_{in} = 0.6R$ and $R_{out} = 1.4R$ are considered, where R is the radius of the air holes.

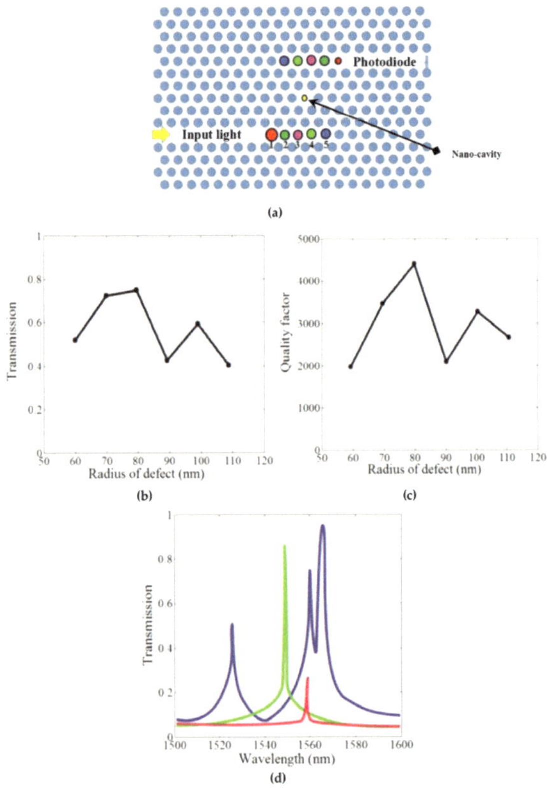

Figure 3. (a) The basic structure of the biosensor, (b) the transmission spectrum, (c) the quality factor for different sizes for nano-cavity, and (d) the transmission spectrum in three states for coupling distance [14].

Also, in this structure, the distance between the waveguides and nano-cavity has been investigated. In **Figure 3(d)**, the transmission spectrum in three states with one (blue curve), two (green curve), and three (red curve) rows of holes as coupling distance is depicted. In the case of one row for coupling distance, the disturb mode is appeared in transmission spectra. If the third row of holes is between the waveguides and the resonator, the intensity of spectrum transmission will be a sharp reduction. But with two rows of holes as distance coupling, the intensity transmission spectrum is desirable and the sensor presents the high quality factor. For a good coupling between the waveguide and nano-cavity, choosing the length of waveguides is important. The length is studied at five states. In each case, the transmission spectra are measured. The output transmission spectra at the five cases can be observed in **Figure 4(a)**. Choosing the three, four, and five states are not suitable and the quality factor of the state 1 is more comparing to the quality factor of state 2. So the state 1 is intended to select the length of the waveguides. **Figure 4(b)** shows the transmission spectrum of the biosensor by changing the refractive index in the range of 1.00–3.00. The resonant wavelength is shifted to the longer wavelengths by increasing the refractive index. The parameters in sensing application include the quality factor, sensitivity, and regression coefficient that are, respectively, obtained as 5248, 55 nm/RIU, and 0.99923.

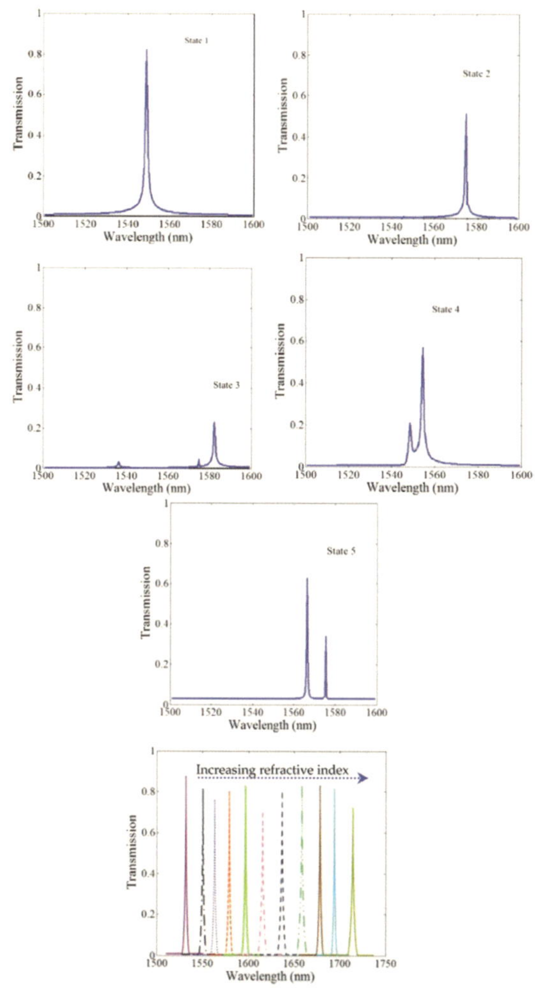

Figure 4. (a) The output transmission obtained by waveguides' length variation for five states and (b) the transmission spectrum of biosensor for some refractive indexes in the range of 1.00–3.00 [14].

4.1.2. Biosensor-based photonic crystal micro-resonator to detect small changes in the refractive index

In 2003, the biosensor based on the photonic crystal micro-resonator was introduced to detect small changes in the refractive index [15]. The sensor was suitable to detect many chemical or biochemical molecules. The transmission spectra of the sensor have a resonance wavelength within the wavelength range of 1.5 μm and the quality factor was equal to 400. The structure was responsive to changes in the refractive index in the range of 1–1.5. The micro-resonator was formed by decreasing the size of the air hole in the center of the structure. The general layout of the structure is represented in **Figure 5(a)**. By changing the refractive index as 1.446, 1.448, 1.450, 1.452, and 1.454, the resonant peak is shifted to about 1500.2 nm, 1500.8 nm, 1501.2 nm, 1501.7 nm, and 1502.1 nm, respectively, as shown in **Figure 5(b)**.

4.1.3. Photonic crystal biosensor based on waveguide and nano-cavity

A new sensor for sensing DNA molecules has been investigated in 2016 by Mohsenirad et al. [16]. The structure of this biosensor consists of hexagonal lattice of air holes in a silicon slab and its operation based on two waveguides and nano-cavity that have been created in this lattice. Pulse is applied to waveguide 1, then resonant mode of resonator is excited, and output signal is recorded at the end of waveguide 2. In recorded transmission spectra, the result of interaction between trapped light and analyte into sensing hole is appeared in the form of resonant peak shift. As shown in **Figure 6(a)**, nano-resonator has been created by reducing the radius of an air hole.

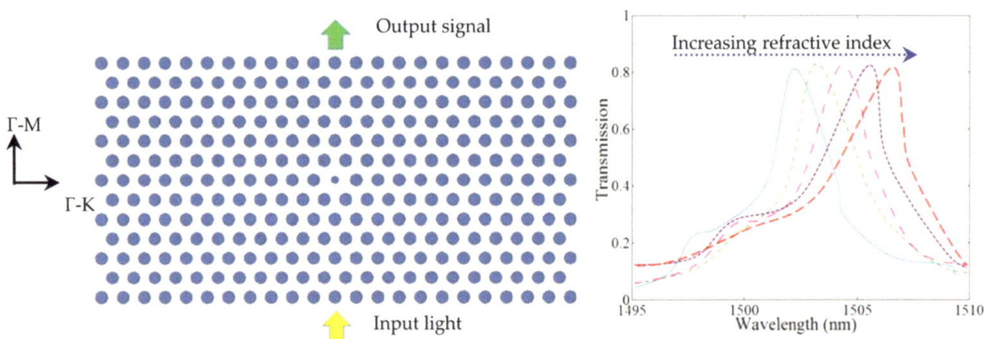

Figure 5. (a) The layout of photonic crystal structure of the biosensor based on micro-resonator to detect small changes in the refractive index and (b) the transmission spectrum of structure for binding biochemical with different refractive indexes [15].

In this structure, physical parameters such as position of resonator, length of the waveguides, radii of the defect, and coupling distance have been investigated in order to achieve an optimized sensor for biochemical sensing applications. In order to choice the best size for nano-resonator, the radius of defect has been studied in the range of 80–100 nm that according to **Figure 6(b)**, the radius of 95 nm has been selected. By increasing the size of sensing hole, sensitivity increases; thus, larger radius is desirable. But quality factor as another important parameter for biosensing applications is contrast with sensitivity.

http://dx.doi.org/ 10.5772/63288

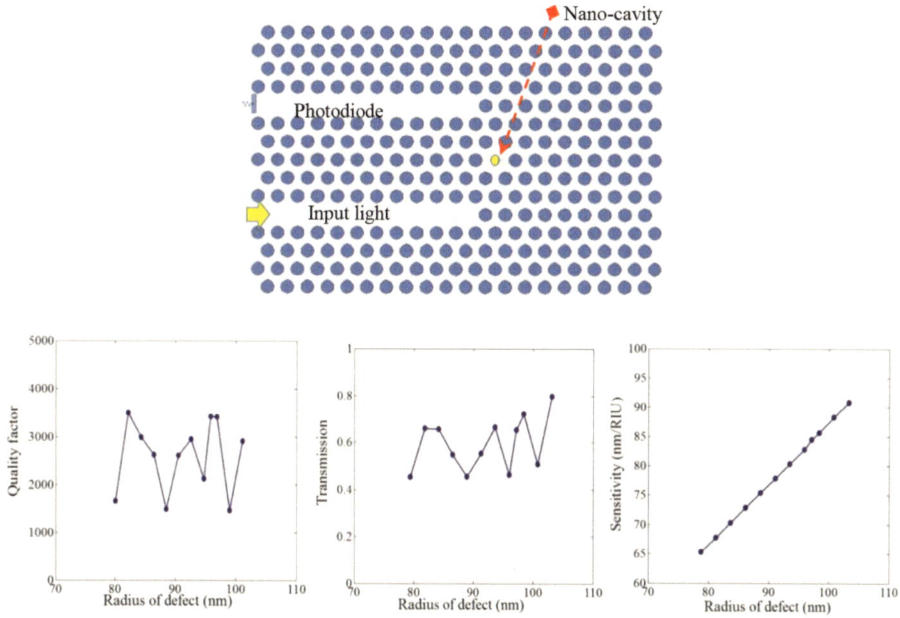

Figure 6. (a) The schematic of photonic crystal biosensor and (b) the sensing parameters of PhC sensor [16].

Since the cavity position affects the sensing parameters of biosensor, transmission spectra of structure for different positions (holes H_1, H_2, and H_3) as are shown in **Figure 7(a)** have been investigated that according to simulation results, H_3 has been selected as the best position.

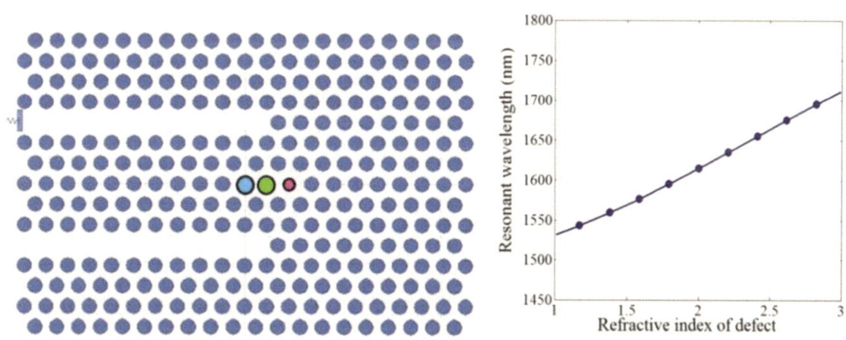

Figure 7. (a) The position of the sensing hole for the nano-resonator (H_1, H_2, and H_3 are, respectively, marked by blue, green, and red holes) and (b) the quasi-linear relation between resonant wavelength and refractive index of defect [16].

Another parameter that in this design has been studied is the length of the waveguide. According to the investigations, when waveguides have been created by eliminating more than 11 holes, spurious modes appeared. The best result for waveguide length is 11 holes. In this design, two rows of air holes have been selected as coupling distance, because by selecting one period as coupling distance, quality factor decreases and spurious mode appears. Also in increment of coupling distance of three periods, light coupling gets difficult and sensitivity decreases.

This biosensor is able to measure a wide range of refractive index of 1.00–3.00 with sensitivity of 83.75 nm/RIU and quality factor of 3051. Also this sensor has quasi-linear operation with correlation coefficient of 0.99954 that is shown in **Figure 7(b)**.

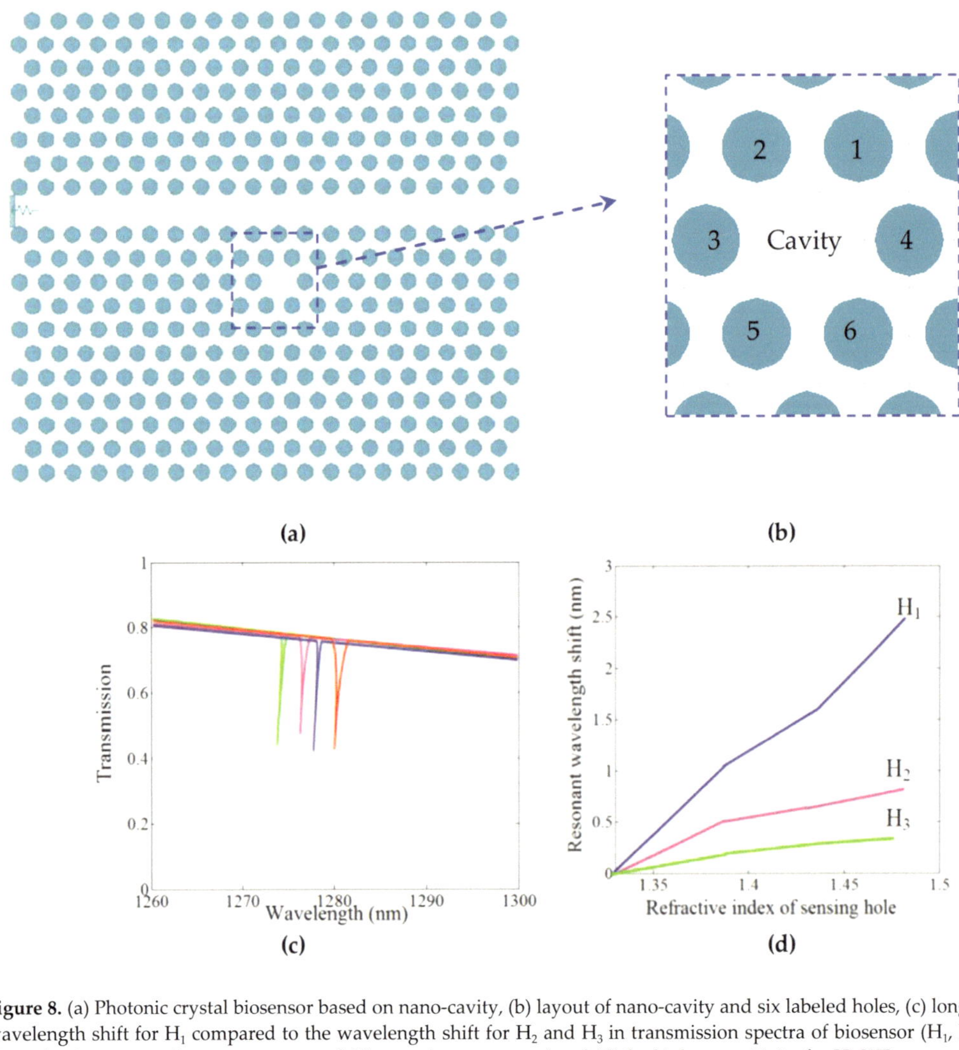

Figure 8. (a) Photonic crystal biosensor based on nano-cavity, (b) layout of nano-cavity and six labeled holes, (c) longer wavelength shift for H_1 compared to the wavelength shift for H_2 and H_3 in transmission spectra of biosensor (H_1, H_2, H_3, and reference are shown as orange, red, blue, and green curves) and (d) the higher sensitivity for H_1 [17].

4.1.4. Biosensor based on nano-resonator

In the biosensor based on nano-resonator structure, the resonator has been created by omitting an air hole and selecting an air hole adjacent to it as the best sensing hole [17]. In this sensor, a waveguide has been considered for applying optical pulse and recording output signal. In

Figure 8(a), nano-cavity and waveguide or in other words point and line defects are shown. Sensing mechanism is based on resonant wavelength shift that this shift caused by the binding biochemical to bioreceptor.

The best position of the sensing hole and coupling distance for obtaining optimized sensor parameters with high quality factor and sensitivity are investigated. For this purpose, six holes around the cavity labeled H_1, H_2, H_3, H_4, H_5, and H_6 are considered as shown in **Figure 8(b)**. Any of these holes was considered as a sensing hole and their effective refractive index variation due to injecting biochemical was investigated.

In **Table 1**, quality factor values and wavelength shift for any hole as a sensing hole are listed. According to **Table 1**, H_1, H_2, and H_3 are roughly equal to H_6, H_5, and H_4, respectively, because of symmetry. By choosing H_1 as sensing hole, resonant wavelength shifts more than other cases, as shown in **Figure 8(c)**. Thus, higher sensitivity is obtained by H_1 as a sensing hole, as shown in **Figure 8(d)**. Among all holes, H_1 has been selected for sensing hole because of its high quality factor and wavelength shift.

Sensing hole	Wavelength shift (nm)	Quality factor
H_1	2.2	4267.67
H_2	0.4	4261.67
H_3	0.9	2558.00
H_4	0.9	2558.00
H_5	0.4	4261.67
H_6	2.2	3200.75

Table 1. List of resonance wavelength shift and quality factor for different sensing holes [17].

Also, the effect of coupling distance has been studied and two rows of air holes were selected as coupling distance between resonator and waveguide. Investigations indicated that with increment of distance, quality factor is increased but sensitivity and output intensity are decreased.

Sensitivity of this sensor is calculated as 165.45 nm/RIU for the refractive indices of 1.00–1.33. Quality factor and detection limit of sensor are ~4000 and 1.2×10^{-3} RIU, respectively.

4.1.5. Multichannel photonic crystal biosensor

Multichannel photonic crystal biosensor consists of some cavities and a waveguide that any cavity has separate operation. This sensor with three cavities is shown in **Figure 9(a)** [17]. The biosensor is able to sense three specific refractive indices simultaneously. In this design, in order to obtain resonators with different resonant wavelengths, the sensing holes have been shifted. By displacing marked hole in **Figure 9(a)**, resonant wavelength shifts, and therefore, new resonant wavelength is appeared as shown in **Figure 9(b)**.

Channels 2 and 3 have been created by deleting an air hole and displacing the sensing hole equal to 0.2 μm and 0.3 μm, respectively. Channel 1 has been obtained by deleting a hole without any displacement. The resonant wavelengths of channels 1, 2, and 3 are 1281.1 nm, 1286.6 nm, and 1296.1 nm, respectively, as shown in **Figure 9(c)**.

Figure 9(d) illustrates the separated operation of channels of multichannel biosensor. In **Figure 9(d)**, refractive index of channel 3 has been changed from 1.00 to 1.45, while refractive index of two other channels has not been changed. The resonant wavelength of channel 3 has been shifted to longer wavelengths, but resonant wavelength of channels 2 and 3 has remained stable. Sensitivity and detection limit of channel 3 have been calculated as S = 1.185 nm/RIU and 0.084 fg, respectively (assuming the injection DNA molecules to sensing hole).

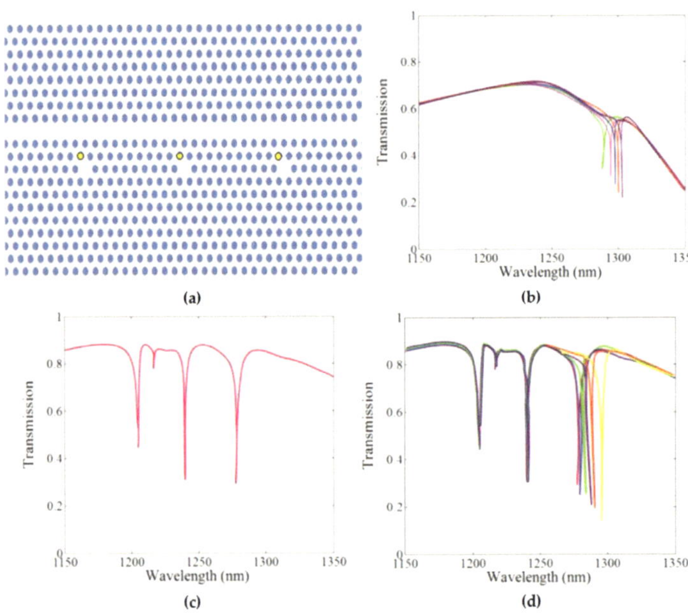

Figure 9. (a) The structure of multichannel biosensor, (b) the effect of sensing hole on transmission spectra, (c) the transmission spectra of multichannel biosensor without any analyte, and (d) transmission spectra of multichannel biosensor for injecting DNA molecules to channel 3 [17].

4.2. Biosensor or biochemical sensor based on photonic crystal LX resonator

In the biosensors and biochemical sensors, another resonator that has been introduced and used is LX resonators. This resonator is formed by removal of the successive air holes of structure which X represents the number of removed holes. In other words, X is the resonator length. For example, L3 resonator is obtained by removing three air holes in the photonic crystal structure. Some of the resonators, namely, L3, L4, L7, and L13, have been provided [18–22]. Some of these sensors based on LX resonators are given in **Figure 10**.

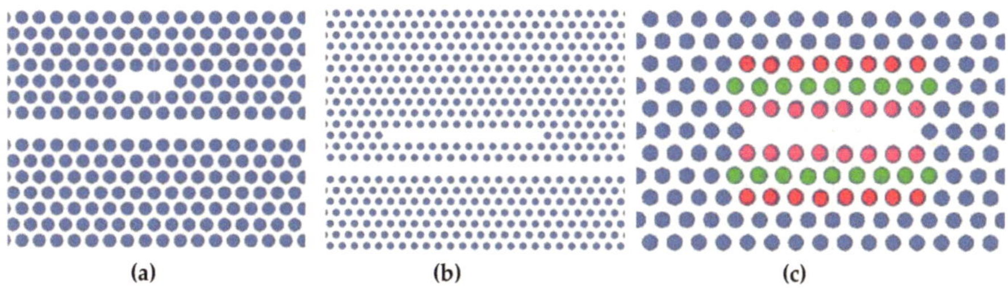

Figure 10. Photonic crystal LX resonators that are suitable for biosensor: (a) L3, (b) L13, and (c) L7 resonators [18–22].

The photonic crystal biochemical sensor based on L4 resonator has been developed in 2005. It was a biochemical sensor for detection of cations and anions [19]. The resonance wavelength of sensor was around 1.5 and the quality factor was obtained about 300–800.

According to conducted research on L3 resonator, Liu and Salemink were reported L7-resonator structure for biosensor applications [20]. The basic parameters of photonic crystal structure such as the lattice constant, the radius of the hole, and the thickness of the silicon slab were selected in such a way that the wavelength was near 1550 nm. The L7 resonator by setting several small holes has been amended. The quality factor of the sensor was 2600. Comparison of simulation results of binding ethanol and water showed that a change in the refractive index equal to 0.03 leads to a change of 12.5 nm in resonance wavelength of transmission. Therefore, a high sensitivity sensor can be created.

Photonic crystal biosensor based on L13 resonator had a high quality factor equal to 9300 [21, 22]. This biosensor has been used to detect protein.

4.3. Biosensor or biochemical sensor based on photonic crystal ring resonator

One of the photonic crystal structures that are used for biosensor applications is the ring resonator [8, 23–26]. For increasing the quality factor and sensitivity of the sensor, reduction of the radius of the ring can be very important. The sensing mechanism of this biosensor or biochemical sensor is based on the measurement of the refractive index change. In continuation, some examples of the biosensor or biochemical sensor based on photonic crystal ring resonator have been presented.

4.3.1. Biosensor based on photonic crystal diamond-shaped nano-ring resonator

Recently, a diamond-shaped nano-ring resonator for biochemical sensing applications has been investigated [23]. Sensing mechanism of sensor is based on RI change mechanism of label-free detection. By binding biochemical molecules to sensing hole, the refractive index will be changed. Thus, the resonance wavelength of the transmission spectrum shifts to longer wavelengths. Defects into the photonic crystal structure are formed by reduction of radius of air holes and are included of a diamond-shaped ring resonator and two end waveguides. The size of ring diameter is equal to three rows of holes. Sketch of the biosensor is shown in **Figure 11(a)**.

In design of the structure, several parameters including reduction of defect holes radius, the end point of the waveguides, and the best holes for attaching chemical or biochemical molecules have been studied [24]. Selecting the best size for the air holes in defect, the R_w/R_a (ratio of the size of the holes of the defect to the size of the air holes) in the range of 0.26–0.39, has been altered. In any case, the quality factor and the intensity of transmission spectra have been measured. According to the results shown in **Figure 11(b)** and **(c)**, the size of 40.8 nm is selected.

For the best point to finish waveguides, five cases have been investigated. The five point cases are marked in **Figure 11(a)**. By comparing all simulation results, the end points, namely, "in ew 4" and "out ew 4," are the best holes to end the waveguide (see **Figure 11(d)**).

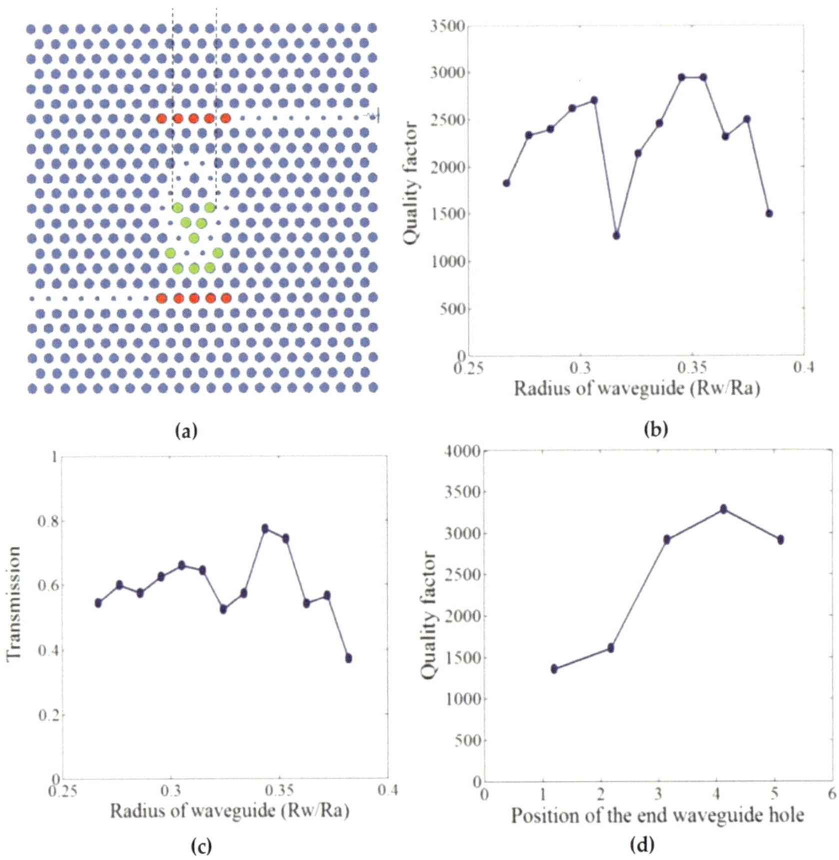

Figure 11. (a) Sketch of the biosensor based on photonic crystal diamond-shaped ring resonator for biochemical sensing applications where the red and green holes indicate position of ew and SHx holes, respectively, (b) the quality factor with respect to the radius of waveguide, (c) the intensity of transmission spectrum, and (d) the quality factor for various end-point waveguides [24].

Also, 10 air holes around the ring resonator for binding biochemical molecules have been studied. The 10 holes, namely, SHx, are shown in **Figure 11(a)**. After attaching the DNA molecule into each holes, the resonance wavelength shift and quality factor are measured. The results have been presented in **Table 2**. Based on the results, the SH_1 can be selected as the sensing hole.

	Sensing hole	Resonant wavelength (nm)	Resonant wavelength shift (nm)	Quality factor
$n = 1$	REF (Reference)	1468.0	–	3716
$n = 1.45$	SH_1	1473.1	5.1	3683
	SH_2	1468.6	0.6	1538
	SH_3	1468.6	0.6	3497
	SH_4	1468.6	0.6	3338
	SH_5	1468.6	0.6	3671
	SH_6	1469.9	1.9	3266
	SH_7	1469.5	1.5	1427
	SH_8	1469.9	1.9	3674
	SH_9	1469.3	1.3	3498
	SH_{10}	1469.9	1.9	3674

Table 2. List of resonance wavelength shift and quality factor for different sensing holes [24].

A relatively large range of refractive index in the sensor from 1.33 to 1.54 is detected. The quality factor of the sensor is about 3700 and the sensitivity of biosensor for DNA detection is 3.4 nm/fg.

4.3.2. *Two-curve-shaped (TCS) biosensor for detecting small refractive index variations*

In some identification of chemical or biochemical molecules, a sensor is required to detect small refractive index changes. A novel design of TCS biosensor has been reported in [25]. The biosensor is constructed based on photonic crystal using the hexagonal lattice of air holes in dielectric slap (see **Figure 12(a)**). Defects into the structure consist of a ring resonator and two waveguides. The defects are configured by removing one or several of air holes. The coupling distance is equal to two rows of air holes. Sensing mechanism of TCS biosensor is based on the IV scheme. By filling sensing hole with chemical or biochemical molecules, the intensity of the transmission spectrum is shifted to lower values.

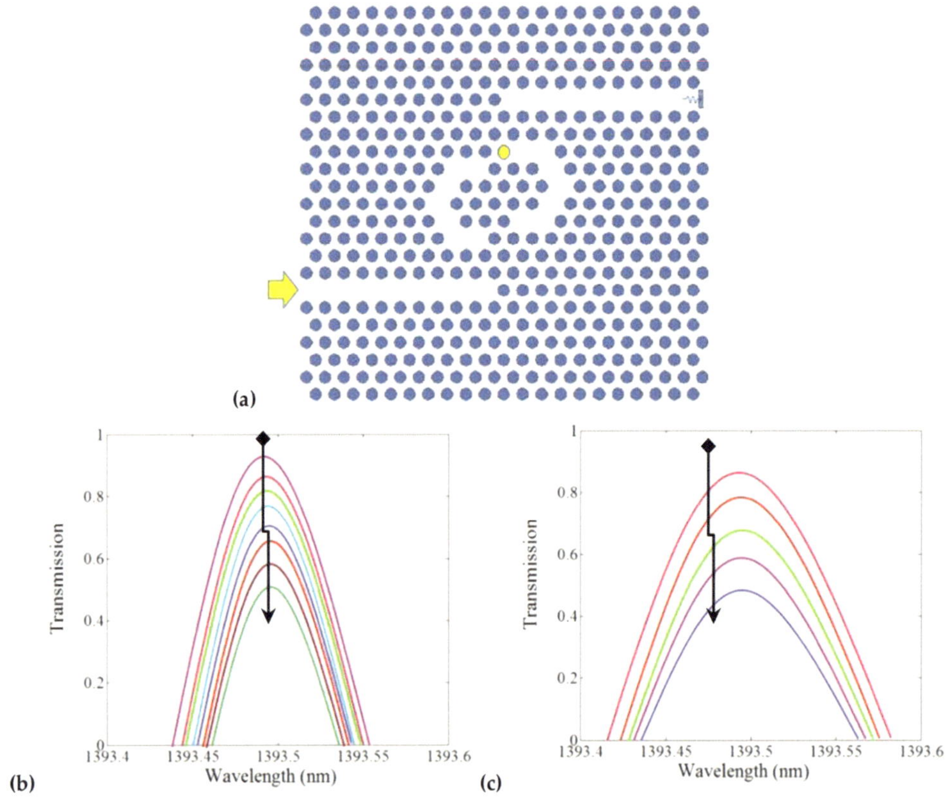

Figure 12. (a) TCS biosensor based on photonic crystal nano-ring resonator (the sensing hole is marked on the structure as yellow hole) [25] and the resonant wavelength of the TCS biosensor when the sensing hole is filled by various biochemical molecules, (b) salt levels of sea water, and (c) glucose concentration (increasing salt level and glucose concentration are shown by arrows) [8].

The present sensor can be used for the detection of the sea water salinity and identification of the glucose concentration [8]. The transmission spectrum when sensing hole plugging by sea water with different salinities is presented in **Figure 12(b)**. Also, the transmission spectrum is shown in **Figure 12(c)** for different percentages of glucose solution attached to the sensing hole. According to **Figures 12(b)** and **12(c)**, in the both cases, the intensity is reduced by increasing concentration.

4.3.3. A novel photonic crystal nano-ring resonator for biochemical sensor applications

Another structure of biochemical sensor is presented that is based on nano-ring resonator shaped by two consecutive curves. One of the curves is made by removing seven air holes [26]. A waveguide to apply light is located at the bottom of the structure. The light from top waveguide is detected by a photodetector. The layout of the biochemical sensor is demonstrated in **Figure 13**. According to **Figure 13**, the coupling distance is considered equal to two rows.

The number of rows between the two curves is expressed in horizontal distance. To have a suitable horizontal distance, this is studied in four states. In the each case, the quality factor and intensity of spectrum transmission are calculated, as shown in **Figure 13(b)** and **(c)**.

The vertical distance meant that curve 2 is several rows higher than curve 1. To examine the vertical distance, the distance from zero to six rows of holes is selected. **Figure 13(d)** and **(e)** presents the results of these modes including the quality factor and the intensity of spectrum transmission.

To check the suitability of the structure for sensing, the refractive index of sensing hole is changed in the range of 1.33–1.48. The normalized curve is shown in **Figure 13(f)**. This curve demonstrates a relatively linear relationship between changes in the refractive index and the intensity shift of transmission spectra. Regression coefficient of the biosensor is obtained about 0.99967. By a unit change in the refractive index of the biochemical sensor, the intensity of transmission spectrum is reduced to 14.26 units.

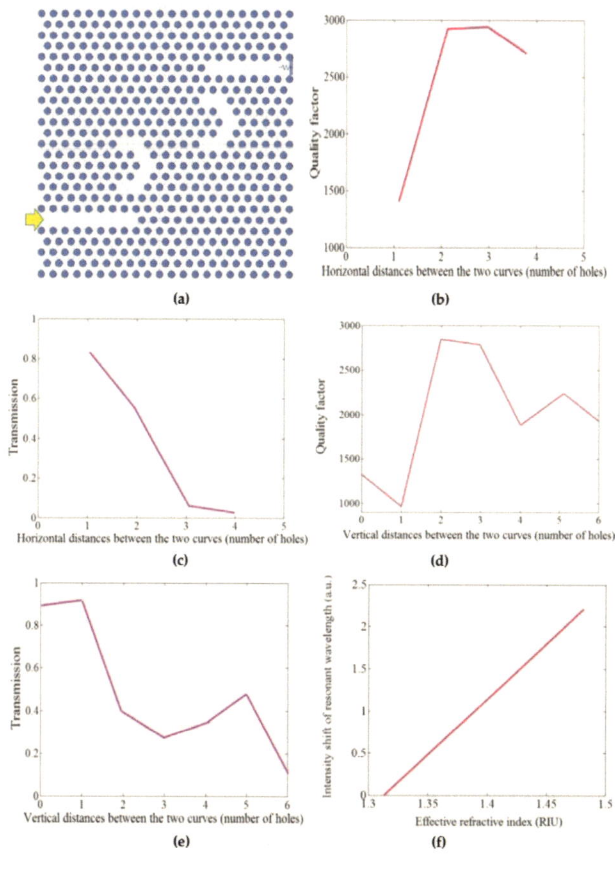

Figure 13. (a) The preliminary layout of the biochemical sensor, (b) the quality factors and (f) the intensity of the resonance wavelength in various horizontal distances between the two curves, (c) the quality factors and (d) the intensity of the resonance wavelength in various vertical distances between the two curves, and (e) the intensity shift of resonant wavelength with respect to the effective refractive index [26].

5. Conclusion

In this chapter, photonic crystals and detection methods in chemical sensors have been described. Also, the most important parameters in sensing applications, such as quality factor, detection limit, and sensitivity, have been introduced. In addition, several sensors based on photonic crystal structures for using chemical or biochemical sensing were investigated. The photonic crystal structures used in presented sensors included resonators or nano-ring resonators. Most of them were based on resonant wavelength shift and the rest were based on the intensity variation scheme. Due to the positive indicators for photonic crystal sensors, they can be appropriate structures for nanotechnology research, food processing, pharmaceutical industry, biology, and environmental applications.

Author details

Saeed Olyaee*, Hamideh Mohsenirad and Ahmad Mohebzadeh-Bahabady

*Address all correspondence to: s_olyaee@srttu.edu

Nano-photonics and Optoelectronics Research Laboratory (NORLab.), Faculty of Electrical Engineering, Shahid Rajaee Teacher Training University (SRTTU), Lavizan, Tehran, Iran

References

[1] J. D. Joannopoulos, S. G. Johnson, J. N. Winn, R. D. Meade, "Photonic crystals molding the flow of light", Princeton University Press, 2007, ISBN: 978-0-691-12456-8.

[2] X. Fan, I. M. White, S. I. Shopova, H. Zhu, J. D. Suter, Y. Sun, "Sensitive optical biosensors for unlabeled targets: A review", Analytica Chimica Acta 620, pp. 8–26, 2008.

[3] S. Olyaee, A. Naraghi, "Design and optimization of index-guiding photonic crystal fiber gas sensor", Photonic Sensors 3, pp. 131–136, 2013.

[4] S. Olyaee, A. Naraghi, V. Ahmadi, "High sensitivity evanescent-field gas sensor based on modified photonic crystal fiber for gas condensate and air pollution monitoring", Optik 125, pp. 596–600, 2014.

[5] S. Olyaee, A. A. Dehghani, "High resolution and wide dynamic range pressure sensor based on two-dimensional photonic crystal", Photonic Sensors 2, pp. 92–96, 2012.

[6] S. Olyaee, A. A. Dehghani, "Ultrasensitive pressure sensor based on point defect resonant cavity in photonic crystal", Sensor Letters 11(10), pp. 1854–1859, 2013.

[7] S. Olyaee, M. Azizi, "Micro-displacement sensor based on high sensitivity photonic crystal", Photonic Sensors 4(3), pp. 220–224, 2014.

[8] S. Olyaee, A. Mohebzadeh-Bahabady, "Two-curve-shaped biosensor for detecting glucose concentration and salinity of seawater based on photonic crystal nano-ring resonator", Sensor Letters 13, pp. 1–4, 2015.

[9] C. Y. Chao, L. J. Guo, Design and optimization of microring resonators in biochemical sensing applications, Journal of Lightwave Technology 24, pp. 1395–1402, 2006.

[10] M. T. Myaing, J. Y. Ye, T. B. Norris, "Enhanced two-photon biosensing with double-clad photonic crystal fibers", Optics Letters 28(14), pp. 1224–1226, 2003.

[11] E. Coscelli, M. Sozzi, F. Poli, D. Passaro, A. Cucinotta, S. Selleri, R. Corradini, R. Marchelli, "Toward a highly specific DNA biosensor: PNA-modified suspended-core photonic crystal fibers", IEEE Journal of Selected Topics in Quantum Electronics 16(4), pp. 967–972, 2010.

[12] S. C. Buswell, V. A. Wright, J. M. Buriak, V. Van, S. Evoy, "Specific detection of proteins using photonic crystal waveguides", Optics Express 16(20), pp. 15949–15957, 2008.

[13] M. G. Scullion, A. Di Falco, T. F. Krauss, "Slotted photonic crystal cavities with integrated microfluidics for biosensing applications", Biosensors and Bioelectronics 27, pp. 101– 105, 2011.

[14] S. Olyaee, S. Najafgholinezhad, "A high quality factor and wide measurement range biosensor based on photonic crystal nanocavity resonator", Sensor Letters 11, pp. 483–488, 2013.

[15] E. Chow, A. Grot, L. W. Mirkarimi, M. Sigalas, G. Girolami, "Ultracompact biochemical sensor built with two-dimensional photonic crystal microcavity", Optics Letters 29(10), pp. 1093–1095, 2004.

[16] H. Mohsenirad, S. Olyaee, M. Seifouri, "Design of a new two-dimensional optical biosensor using photonic crystal waveguides and a nanocavity", Photonics & Lasers in Medicine 5(1), 2016.

[17] S. Olyaee, S. Najafgholinezhad, "Computational study of a label-free biosensor based on a photonic crystal nanocavity resonator", Applied Optics 52(29), pp. 7206–7213, 2013.

[18] D. Dorfner, T. Zabel, T. Hürlimann, N. Hauke, L. Frandsen, U. Rant, G. Abstreiter, J. Finley, "Photonic crystal nanostructures for optical biosensing applications", Biosensors and Bioelectronics 24, pp. 3688–3692, 2009.

[19] S. Chakravarty, J. Topolanik, P. Bhattacharya, S. Chakrabarti, Y. Kang, M. E. Meyerhoff, "Ion detection with photonic crystal microcavities", Optics Letters 30(19), pp. 2578–2580, 2005.

[20] Y. Liu, H. W. M. Salemink, "Photonic crystal-based all-optical on-chip sensor", Optics Express 20(18), pp. 19912–19920, 2012.

[21] S. Chakravarty, W. C. Lai, Y. Zou, H. A. Drabkin, R. M. Gemmill, G. Simon, S. Chin, R. Chen, "Multiplexed specific label-free detection of NCI-H358 lung cancer cell line lysates with silicon based photonic crystal microcavity biosensors", Biosensors and Bioelectronics 43, pp. 50–55, 2013.

[22] S. Chakravarty, Y. Zou, W. C. Lai, R. Chen, "Slow light engineering for high Q high sensitivity photonic crystal micro cavity bio-sensors in silicon", Biosensors and Bioelectronics 38, pp. 170–176, 2012.

[23] S. Olyaee, A. Mohebzadeh-Bahabady, "A diamond-shaped bio-sensor based on two-dimensional photonic crystal nano-ring resonator", IEEE, IET 10th International Symposium on Communication Systems, Networks and Digital Signal Processing (CSNDSP 2014), 23–25 July 2014.

[24] S. Olyaee, A. Mohebzadeh-Bahabady, "Design and optimization of diamond-shaped biosensor using photonic crystal nano-ring resonator", Optik 126, pp. 2560–2564, 2015.

[25] S. Olyaee, A. Mohebzadeh-Bahabady, "Two-curve-shaped biosensor using photonic crystal nano-ring resonators", Journal of Nanostructures 4, pp. 303–308, 2014.

[26] S. Olyaee, A. Mohebzadeh-Bahabady, "Designing a novel photonic crystal nano-ring resonator for biosensor application", Optical and Quantum Electronics 47(7), pp. 1881–1888, 2015.

Section 3

Semiconductor Chemical Sensor

Chapter 4

Perovskites-Based Nanomaterials for Chemical Sensors

Morteza Enhessari and Ali Salehabadi

Additional information is available at the end of the chapter

http://dx.doi.org/10.5772/62559

Abstract

The perovskite structure is adopted by many compounds in solid-state chemistry. The sensitivity, selectivity, and stability of many perovskite nanomaterials have been devoted the most attention for chemical sensors. They are capable to sense the level of small molecules such as O_2, NO, and CO. This chapter provides a comprehensive overview of perovskite nanoscale materials that concentrate on chemical sensors. The perovskite structure, with two differently sized cations, is amenable to a variety of dopant additions. This flexibility allows for the control of transport and catalytic properties, which are important for improving sensor performance. We devote the most attention on the synthesis, structural information, and sensing mechanism. We will later elaborate on the development mechanism of chemical sensors based on perovskite nanomaterials. We conclude this chapter with the personal perspectives on the directions toward future works on a novelty of nanostructured chemical sensors.

Keywords: Perovskite, Nanomaterial, Chemical sensor, Gas sensing material, Semiconductor

1. Introduction

The 3D-metal monooxides such as MnO, FeO, CoO, and NiO are semiconductors with low electrical conductivities. In general, the as-mentioned metal oxides have such large band gaps that they are insulators. The structure with general formula ABX_3, containing 12-coordinate hole of BX_3 is occupied by a large A ion (A: Ni, Pb, Fe, Co, Zn, etc) are known as perovskites. The tetragonal structure of perovskites showing the local displacements leads to the variation in electric behavior of this material (**Figure 1**).

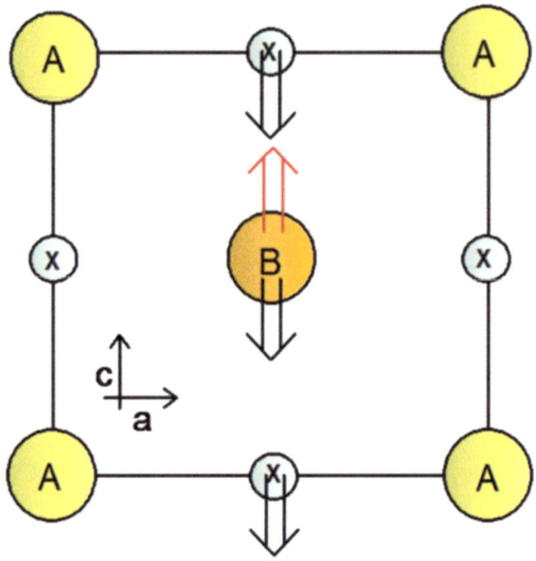

Figure 1. The local ion displacements of a tetragonal ABX_3 structure [1].

Many applications of perovskites (here ABO_3) are reported in electrodes of solid oxide fuel cells [2], metal–air batteries [3], gas sensors [4], and high-performance catalysts [5].

Metal-oxide semiconductors, such as semiconducting tin oxide (SnO_2), have been used as chemical sensors for a number of years. Applications include environmental monitoring, fire detection, and vehicle monitoring [4]. The fundamental sensing mechanism of these metal-oxide-based gas sensors relies upon the change in electrical conductivity due to the interaction between the gases in the environment and oxygen in the grain boundaries.

A chemical sensor is a device that transforms chemical information into an analytically useful signal. The chemical information may originate from a chemical reaction of the analyte or from a physical property of the investigated system. They have applications in different areas such as medicine, home safety, environmental pollution, and many others. Chemical sensors are classified according to the operating principle of the transducer (**Table 1**).

A variety of chemical methods have been reported for the synthesis of semiconducting materials including perovskites, spinels, or illminites which include ball-milling [6, 7], combustion synthesis [8–10], co-precipitation [11], sol–gel [12–15], radio frequency sputtering [16–19], solid-state reaction [20–22], and molten-salt method [23, 24].

In this chapter, we will review the routine reliable synthesis method, structural information, and gas sensing behavior of (mixed) metal-oxide semiconductors with tunable functionalities used in electrical devices for chemical sensors. Novel functional nanostructure semiconductors will be presented.

Type of chemical sensor	Source	Example
Oprical (optodes)	Optical phenomena	Absorbance, reflectance Luminescence Fluorescence Opto-thermal effect Light scattering
Electro chemical	Electrochemical process	Voltammetric sensors Potentiometric sensors Chemically sensitized field effect transistor potentiometric solid electrolyte gas sensors
Electrical	Electrical properties	Metal-oxide semiconductor Organic semiconductor Electrolytic conductivity Electric permittivity
Mass sensitive	Mass change at a specially modified surface	Piezoelectric devices Surface acoustic wave devices
Magnetic	Change of paramagnetic properties	Oxygen monitors
Thermometric	Heat effects of a specific chemical reaction or adsorption	Combustion reaction Enzymatic reaction Optothermal device
Radiation	Change of Physical properties	X-, p- or r-radiation Chemical composition

Table 1. Classification of chemical sensors

1.1. Semiconductors chemistry

Semiconductors are insulators at absolute zero. Above this temperature and below the melting point of the solid, the metal acts as a semiconductor. Semiconductors are generally classified on the basis of their composition and particle size. Procreation of semiconductors with typical band gaps (a few eV's corresponding to 100–200 kJ mol^{-1}) usually occurs between *p*-block metals and group 13/14 metalloids in combination with *chalcogenides* and *pnictides* [1].

Perovskites are an example of especial class of semiconductors (**Figure 2**). In this structure, the unit cell is not centrosymmetric and the crystal develops a permanent electric polarization as a result of ion displacements (refer to **Figure 1**). Barium titanate, $BaTiO_3$, is one of the interesting examples of perovskite. This compound, at above 120°C, exhibits a cubic structure while a lower symmetry, that is, a tetragonal unit cell at room temperature due to ions displacement. Fe_2TiO_5 [25], $NiTiO_3$ [26, 27], $CoTiO_3$ [15, 28], $BaZrO_3$ [12], La_2CuO_4 [29], $LaMnO_3$ [5], $MnTiO_3$ [14], and $PbTiO_3$ [30] are some important examples of common perovskite oxide semiconductors.

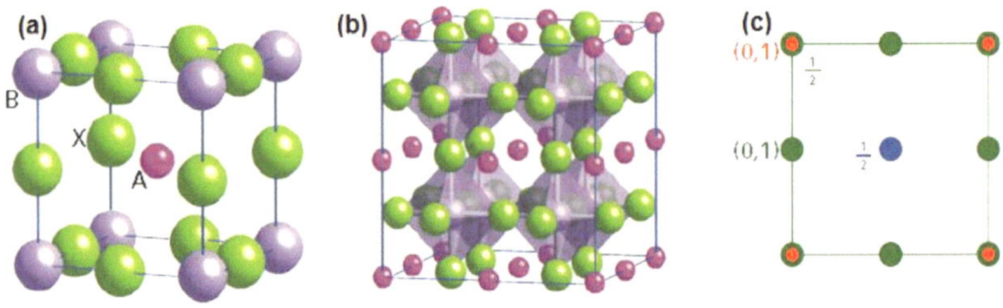

Figure 2. Schematic representation of (a) perovskite (ABX_3) structure, (b) emphasizes of octahedral sites, and (c) its respective projection [1].

1.2. Semiconducting Chemical Sensors

Chemical sensors are generally classified according to the shape of metal oxides to nanotubes, nanorods, nanobelts, and nanowires [16]. A nanotube sensors include metal-oxide tubes such as Co_3O_4 (superior gas sensing capabilities toward H_2), SnO_2 (gas sensor to ethanol), and TiO_2 (hydrogen sensor). Nanorod-based sensors are involved in metal oxides such as ZnO, MoO_3, and tungsten oxide nanosensors, polymer nanorods such as poly(3,4-ethylene-dioxythiophene)-nanosensor, and metal nanorods such as Au nanosensor. As for nanobelt-based sensors, the attentions have been devoted on metal oxides such as ZnO (Co and O_2 sensors), SnO_2 (NO_2 sensors), and V_2O_5 (stable ethanol sensor) nanosensors, especially on ZnO nanobelts sensors. Nanowire metal oxides such as In_2O_3, SnO_2, ZnO, and Ga_2O_3 are used in NO_2 sensor, O_2 and CO sensors, NH_3 sensor, and ethanol sensor, respectively.

Perovskites with a general formula of ABO_3, a typical band gap of 3–4 eV and good thermal stability, are interesting choices for gas sensing materials [6]. A suitable material for a chemical sensor particularly gas sensors must not only have an appropriate electronic structure but also a good affinity with the target reactant which satisfies a number of other requirements such as manufacturability, hydrothermal stability, poisoning resistance, and adaptability with existing technologies. A wide variety of techniques are capable of synthesizing perovskites (nano)- materials. Sol–gel, impregnation, and precipitation are some of the most promising ones from efficiency and scale-up perspectives.

Many researchers have reported the gas sensing devices for dangerous gases such as carbon oxides (CO, CO_2), nitrogen oxides (NO, NO_2), and sulfur dioxide (SO_2) released from high-temperature combustion processes. CH_4 is another potent greenhouse gas that leads to global warming [31–34].

The perovskite oxide generally used in oxygen sensors is $SrTiO_3$ [35]. The oxygen partial pressure dependence of the conductivity of undoped $SrTiO_3$ contains both n-type and p-type regimes.

Gas sensing properties of nanocrystalline perovskite, $LaFeO_3$ [36] and $BaTiO_3$ [37], were successfully involved in previous literatures. It is concluded that the sensor based on the $LaFeO_3$ powders has a considerable sensing response to carbon dioxide. On the other hand, the $BaTiO_3$ sensor has a good selectivity to LPG against CO_2, H_2, NH_3, C_2H_5OH, and Cl_2. The $BaTiO_3$ thin films exhibit rapid response recovery which is one of the main features of this sensor. The defect structure and conduction properties of $BaTiO_3$ are also similar to those of $SrTiO_3$. However, the conductivity of $BaTiO_3$ is higher than that of $SrTiO_3$ in the p-type regime.

Hydrogen-sensitive semiconductor $SrCe_{0.9}Yb_{0.1}O_3$ nanopowders with tunable sensing functionalities toward H_2 and H_2S were synthesized in our group [13]. We observed a noticeable gas sensing behavior of the perovskite at room temperatures with semi-spherical and porous structures of nanoparticles.

High sensitivity and selectivity of mixed potential sensor based on Pt/YSZ/$SmFeO_3$ for NO_2 gas are investigated by Giang et al. [38]. They concluded that the high sensing performances of the as-mentioned sensor to NO_2 could be related to high catalytic active of the sensing material, that is, $SmFeO_3$. The adsorption of exposed gases on the surface of $SmFeO_3$ has been considered the factor that affects on the performance of the sensor Pt/YSZ-8/$SmFeO_3$.

CO sensitive nanocrystalline $LaCoO_3$ and $La_{1-x}Ce_xCoO_3$ perovskite sensor were investigated by Dhivia et al. [6] and Ghasdi et al. [31]. They inferred that the oxygen mobility increased by increasing the surface area. Moreover, the maximum response ratio of CO exhibited a good correlation with the total amount of desorbed oxygen.

Hydrocarbon gas sensing of nanocrystalline perovskite oxides, $LnFeO_3$, shows a superior response to gas sensing characteristics containing methane, propane, and n-hexane. Moreover, the rare earth elements of the as-mentioned perovskite play an important role to gas sensitivities [39].

Nanostructured perovskite, $CdTiO_3$ films for methane sensing, is one of the recent studied examples of semiconductor for chemical sensors [40]. The gas sensing mechanism of $CdTiO_3$ sensor is of surface-controlled type. The variation in the resistance of sensing material caused by the adsorption and desorption of gas molecules such as O_2. The oxygen adsorption on the $CdTiO_3$ film surface depends on the temperature. At high temperatures, O_2 dissociates into atomic oxygen $O^-_{(ads)}$(Eq. 1)

$$\begin{aligned} & O_{2(\mathrm{gas})} \leftrightarrow O_{2(\mathrm{ads})} \\ & O_{2(\mathrm{ads})} + e^- \rightarrow O^-_{2(\mathrm{ads})} \\ & O^-_{2(\mathrm{ads})} + e^- \rightarrow 2O^-_{(\mathrm{ads})} \end{aligned} \quad (1)$$

The chemistry image, the classification of perovskites applicable as gas sensing materials, the fabrications, and the performance mechanism of gas sensors will be discussed in the following sections. It must be noted that in this chapter the perovskites classes are named based on "B" site in general formula of ABO_3, such as stanate ($BaSnO_3$), titanate, and niobate.

2. Classification of perovskites

As mentioned before, the typical chemical formula of the perovskite structure is ABO_3 and AB_2O_4, where A and B denote two different cations. The ilmenite structure has the same composition as the perovskite, that is, ABO_3; however, A and B in this structure are cations of approximately the same size that occupy an octahedral site.

Compound	Lattice parameter/× 10 nm		
	A	B	c
		Cubic structure	
$KTaO_3$	3.989		
$NaTaO_3$	3.929		
$NaNbO_3$	3.949		
$BaMnO_3$	4.040		
$BaZrO_3$	4.193		
$SrTiO_3$	3.904		
$KMnF_3$	4.189		
$KFeF_3$	4.121		
		Tetragonal structure	
$BiAlO_3$	7.61		7.94
$PbSnO_3$	7.86		8.13
$BaTiO_3$	3.994		4.038
$PdTiO_3$	3.899		4.153
$TlMnCl_3$	5.02		5.04
		$LaAlO_3$ type	
$LaAlO_3$	5.357	α=60° 06′	
$LaNiO_3$	5.461	α=60° 05′	
$BiFeO_3$	5.632	α=60° 06′	
$KNbO_3$	4.016	α=60° 06′	
		$GdFeO_3$ type	
$GdFeO_3$	5.346	5.616	7.668
$YFeO_3$	5.283	5.592	7.603
$NdGaO_3$	5.426	5.502	7.706
$CaTiO_3$	5.381	5.443	7.645
$NaMgF_3$	5.363	5.503	7.676

Table 2. Typical perovskite compound [41]

Various combinations of charged cations in the A and B sites of perovskite compounds such as 1 + 5 and 2 + 4 have been reported. However, a complex combinations are also investigated, such as $M(B'_{1/2}B''_{1/2})O_3$, where M can be Pb or La, B′ can be Sc, Fe, Ni, or Mg and B″: Nb, Ta, Ru(IV), or Ir(IV). Typical perovskite compounds with various lattice structures are listed in **Table 2**.

In **Table 3**, the as-reported elements that can be incorporated within the perovskite structure are listed (Self-test: to be completed by the readers). It is obvious that almost all elements except noble gases can be occupied either A or B lattice structure in the perovskite lattice. The crystal structure of perovskites is dependent on the size and the nature of the A and B atoms. Simplifying the possible lattice interaction between elements forms a perovskite structure for gas sensing materials.

B/A	Na	K	Rb	Ca	Sr	Ba	Y	La	Hf	Ag	Cd	Ln	Tl	Pb	Ce	Th	Pr	Nd	Sm	Eu	Gd	Tb	Dy	Ho	Er	Tm	Yb	Lu	Th
Li																													
Mg																													
Sc																													
Ti				✓	✓			✓						✓															
Zr														✓															
V																													
Nb	✓	✓				✓																							
Ta		✓																											
Cr				✓			✓	✓																					
Mo																													
W																													
Mn				✓	✓		✓	✓																					
Tc																													
Re																													
Fe		✓		✓	✓	✓		✓*				✓			✓				✓										
Ru																													
Os																													
Co						✓	✓	✓							✓														
Ni						✓		✓																					
Pd					✓																								
Pt																													
Cu		✓																											
Zn																													
Al																													
Ga																													
In																													
Ge																													
Sn				✓		✓																							
U																													
Np																													
Pu																													

"✓" Represents the available, as-synthesized gas sensing perovskites; "*" corresponds to $LaFeO_3$ [42].

Table 3. (Self-test) elements that can occupy perovskite structure (ABO_3, AB_2O_4)

Another class of perovskites forms by the replacement of B-site elements with a dopant. The as-mentioned structure with general formula $A_3MB_2O_9$ is called as superstructure perovskites. In fact, large differences in ionic radii lead to the formation of this huge structure. Here, M can be Fe, Co, Ni, Zn, or Ca like $Ba_3SrTa_2O_9$. Moreover, a new class of superstructure perovskite is invented where the ordering of cation vacancies located on A sites: for example, MNb_3O_9 (M-La, Ce, Pr, Nb) and MTa_3O_9 (M-La, Ce, Pr, Nd, Sm, Gd, Dy, Ho, Y, Er).

Typical polymorphs of the perovskite structure are brownmillerite ($A_2B_2O_5$). This is oxygen deficient type of perovskite, the oxygen vacancy is ordered [41, 43]. In this structure, as the content of oxygen drops, the phase changes from the cubic perovskite to tetragonal, to orthorhombic, and finally to the brownmillerite structure. These phases are in fact a repeating sequence of octahedral and tetrahedral layers [44]. In the typical example of a non-stoichiometric ternary oxide such as $SrFeO_{2.5+x}$ ($0 < x < 0.5$), there is a series of phases that is derived from the perovskite structure. Stepwise dropping of the oxygen content, the phase changes from the cubic (x = 0.5) to tetragonal (x = 0.35), to orthorhombic (x = 0.25), and to the brownmillerite structure (x = 0) [43, 44]. The as-mentioned phenomenon can be observed clearly in the reaction isotherm profiles (**Figure 3**). In the regions where two phases co-exist, the x-value increases greatly for the small increase in oxygen pressure.

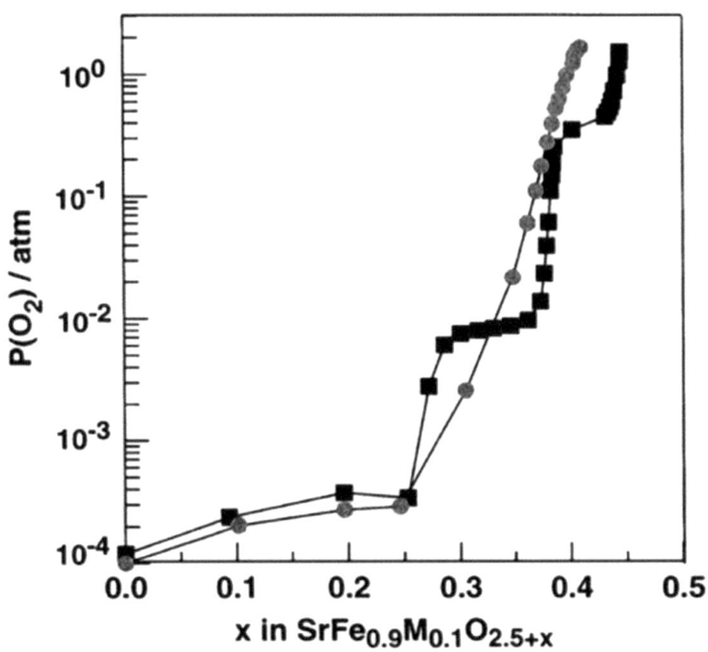

Figure 3. Reaction isotherm of $SrFeO_{2.5+x}$ in a substituted $SrFe_{0.9}Cu_{0.1}O_{2.5}$ [44].

Ruddelsden–Popper compounds are a group of perovskite with the general formula $(ABO_3)_nAO$ (**Figure 4**). Some examples of this structure are $Sr_3Ti_2O_7$ and $Sr_4Ti_3O_{10}$.

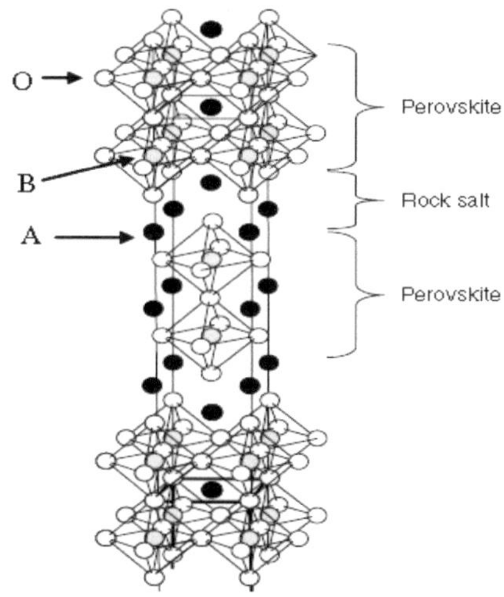

Figure 4. Ruddelsden–popper structure [41].

3. Sensing materials and mechanism

Sensor technology has widely distributed as a basic enabling technology in many instances. Many applications of sensing devices particularly in intelligent manufacturing processing have been reported ranging from assessing the integrity of aircraft to monitoring the environment for hazardous chemicals [45, 46]. The oxygen deficient crystal structure in semiconducting oxide materials is responsible for the change in resistance an oxide sensor [1]. This is due to the adsorbed surface species [41]. The reducing gas develops a shrink charge region which allows better charge movement.

A chemical/gas sensor provides an electrical/optical output in response to chemical and physical interactions with chemicals/gases. In particular, the gas sensors have been used for various industrial and/or safety applications and chemical such as a) determination of gas leakage [41, 47–49], novel LPG sensitive materials working in the entire range of least explosive limit (LEL) to upper explosive limit (UEL) [30–32]. Nanostructured perovskite materials can improve the sensing properties of the sensor due to high surface-to-volume ratio characteristics. Knowledge of nanoscale perovskite sensing materials have been achieved a great attraction to serve as a novel gas sensing materials at the lower working temperatures.

How does a sensing materials work? Band theory is a principle of chemical sensors technology which postulates that when atoms or molecules are aggregated in the solid state, the outer atomic orbitals are split into bonding and antibonding orbitals and mix to form two series of closely spaced energy levels. In general, in a gas sensor, the gas species react with metal-oxide particle surfaces thereby trapping electrons (**Figure 5**). This is the process of chemisorptions. A resistance layer from an electron-depleted space (cloud) charge on the n-type (the majority of charge carriers are electrons) particle surface or conducting layer from accumulated holes on the p-type (the positive holes being the majority of charge carriers) particle [50]. Continued reaction between gas molecules and chemisorbed oxygen causes electrons to release from oxygen back to oxide. The current changes the electrical conductivity in the space charge layer. The conductivity increases in the case of n-type oxides, as a greater charge carrier (electron) concentration, and decreases in p-type oxides, as electrons recombine with holes. Time-by-time changing of conductivity in the space charge layer varies the overall electrical resistance of the oxide. Hence, the space (cloud) charge surrounding the bulk material directly influenced the sensitivity, that is, thicker the space charge layer higher the sensitivity. The change of sensor resistance depends on a type of metal-oxide semiconductors, here perovskites. The changes in sensor resistance upon exposure to the target gas/reducing gas in the cases of n-type and p-type perovskite sensors are reported by Choopon et al. [51] as illustrated in **Figure 6**.

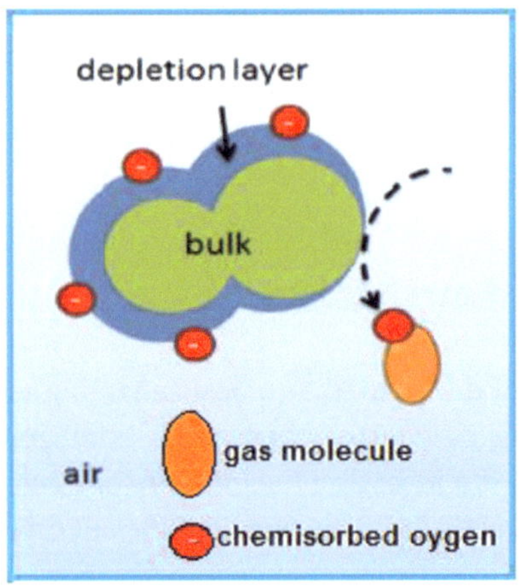

Figure 5. Mechanism of gas sensing [50].

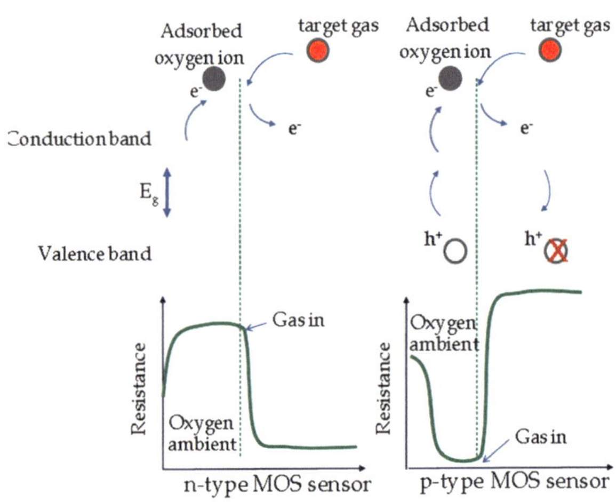

Figure 6. Change of sensor resistance upon exposure to the target gas [51].

4. Application of perovskite nanomaterials in gas sensor

Variety of component and high chemical stability are two important characteristic of perovskite nanomaterials applicable for catalysts in various reactions. It is impractical to consider the classification of perovskites without their applications in sensing devices. The advent of high-performance solid-state gas sensors has motivated several scientists in searching the new perovskite materials and investigating their gas sensing properties. In general, the perovskites are either oxidation catalysts or oxygen-activated catalysts as an alternative to catalyst containing precious metals, or a model for active sites. The stability of the perovskite structure allows the invention of new compounds with an unusual valence state of elements or a high extent of oxygen deficiency [41]. In **Table 4**, we summarize the as-reported perovskite nanomaterials in sensing device. We mentioned that the high catalytic activity of perovskite oxides is based on the high surface activity to oxygen reduction ratio or large number of oxygen vacancies in the particular structure. Among the various catalytic reactions studied, automobile exhaust gas, various pollutant gases such as H_2S, and NH_3, NO_x decomposition reaction gas, hydrogen gas, methanol, and LP gas attract particular attention. The perovskite materials can be used as a thin film (nanocomposites) or nanopowders.

Sensing materials	NO_x	CH_4	NH_3	LPG	C_2H_5OH	Acetone	Refs.
Perovskite							
Titanate		$Sr(Ti_{0.65}Fe_{0.35})O_3$, $Pb(Zr_{0.2}Ti_{0.8})O_3$	$BaTiO_3$, $CdTiO_3$, $Na_2Ti_3O_7$		$Li_{0.35}La_{0.55}TiO_3$		[52–57]
Ferrite	$Cu_xFe_{3-x}O_4$, $Mg_{0.5}Zn_{0.5}Fe_2O_4$	$SrTi_{0.6}Fe_{0.4}O_{3-\delta}$ $BaFeO_3$	$La_{0.8}Sr_{0.2}Fe_{1-x}Cu_xO_3$, $Cu_{0.5}Zn_{0.5}W_{0.3}Fe_{1.7}O_4$	$Co_{1-x}Ni_xFe_2O_4$, $Mg_{0.5}Zn_{0.5}Fe_2O_4$		$NiFe_2O_4$	[17, 58–62]
Cobaltite		$Bi_{10}Co_{16}O_{38}$, $Ln_{1-x}Sr_xCoO_3$		$NdCoO_3$, $Nd_{0.8}Sr_{0.2}CoO_3$	$Bi_{12}(Bi_{0.55}Co_{0.45})O_{19.6}$ $LnBaCo_2O_{5+\delta}$, $Bi_{10}Co_{16}O_{38}$ $Ba_{0.5}Sr_{0.5}Co_{0.8}Fe_{0.2}O_{3-\delta}$		[63–69]
Cobaltate				$YCoO_3$, $LaCoO_3$ $La_{1-x}Ce_xCoO_3$ $NdCoO_3$			[6, 35, 60, 65–68]
Mangenate		$La_{0.8}Sr_{0.2}Al_{0.9}Mn_{0.1}O_3$, $Ln_{1-x}Ca(Sr)_xMnO_3$					[67, 70]
Cerate	$SrCe_{0.95}Yb_{0.05}O_{3-\alpha}$, $SrCeO_3$, $SrCe_{0.95}Tb_{0.05}O_{3-\delta}$		$BaCe_{0.90}Gd_{0.1}O_{3-\delta}$				[9, 71–73]
Niobate	$FeNbO_4$		$BaNbO_3$			$FeNbO_4$	[74, 75]
Nickelate				$LaNi0_3$			[76]
Stanate	$BaSnO_3$, Zn_2SnO_4, $ZnSnO_3$		$Ba_{1-x}Ni_xSnO_3$ $Ba_{1-x}La_xSnO_3$, $ZnSnO_3$	$BaSnO_3$			[77–83]
Zirconate	$PbZrO_3$		$CaZrO_3$				[20, 48, 84]
Chromate			$MgCr_2O_4$				[85, 86]
Molybdate	Bi_2MoO_6	$Bi_3FeMo_2O_{12}$	$ZnMoO_4$ $Bi_3FeO_4(MoO_4)_2$ $NiMoO_4$, $CuMoO_4$, $PbMoO_4$	$Bi_3FeMo_2O_{12}$			[87–89]
Tungstate	Sn_xWO_{3+x}		$ZnWO_4$, $MnWO_4$	$CoWO_4$, Sn_xWO_{3+x}, $SnW0_4$			[90–96]

Sensing materials	**NO_x**	**CH_4**	**NH_3**	**LPG**	**C_2H_5OH**	**Acetone**	**Refs.**
Perovskite							
Sensing materials	NO_x	CH_4	NH_3	LPG	C_2H_5OH	Acetone	Refs.
Perovskite							
Titanate				$ZnTiO_3$	$CoTiO_3$, $Cr_{1.7}Ti_{0.3}O_3$, Zn_2TiO_4,		[97–100]
Ferrite	$LaFeO_3$, $Co_{1-x}Mn_xFe_2O_4$, $CoFe_2O_4$	$Mg_{0.5}Zn_{0.5}Fe_2O_4$	$Ni_xZn_{1-x}Fe_2O_4$	$CuFe_2O_4$, $Ni_{1-x}Co_xFe_2O_4$	$Co_{1-x}Ni_xFe_2O_4$, $ZnFe_2O_4$, $Ni_xZn_{1-x}Fe_2O_4$, $GaFeO_3Ni_{1-x}Co_xFe_2O_4$ $Ni_{1-x}Co_xFe_2O_4$	$CoFe_2O_4$, $NiFe_2O_4$,	[11, 23, 47, 95, 96, 98–101]
Cobaltate	$YCoO_3$	$YCoO_3$					
Mangenate	$LaMnO_{3+\delta}$, $La_{0.6}Ca_{0.4}Mn_{1-x}Ni_xO_3$,						[42, 102]
Niobate		$LiNbO_3$	$CrNbO_4$, $InNbO_4$	$InNbO_4$, $FeNbO_4$	$AlNbO_4$, $CrNbO_4$, $InNbO_4$		[70, 97, 102–107]
Nickelate	$LaNi0_3$						[76]
Stanate	Zn_2SnO_4	Zn_2SnO_4	$CaSnO_3$, $ZnSnO_3$	Zn_2SnO_4, $ZnSnO_3$	$CaSnO_3$, Zn_2SnO_4,		[72–74, 108, 109]
Chromate					$LaCr_{1-x}Ti_xO_3$		[8]
Molybdate			Bi_2MoO_6, $NiMoO_4$		$Bi_3FeMo_2O_{12}$		[110, 111]
Tungstate	$CuWO_4$, $SnW0_4$, $MgWO_4$, $ZnWO_4$, $BaWO_4$		$CoWO_4$				[87–89, 110, 112]

Table 4. Gas sensing perovskite oxides

From **Table 4**, it is obvious that the various combinations of elements are used to produce the perovskite nanomaterials for the as-mentioned sensing devices. Among them, the humidity, LPG, ethanol, and pollution gases have been achieved a great attention.

Humidity measurement is one of the most significant issues in various areas of applications such as instrumentation, automated systems, agriculture, climatology, and GIS. Molybdate- , stanate- , and titanate-based perovskites are three powerful groups of materials used in this criterion. When perovskite oxides are exposed to moisture, interaction occurs in three stages: (a) a few water vapor molecules are chemisorbed at the neck of the crystalline grains on activated sites of the surface and form hydroxyl groups. As a result of this interaction, the hydroxyl group of each water molecule is adsorbed on metal cations and possesses high charge carrier density and strong electrostatic fields, and producing mobile protons. The protons migrate on the surface and react with the neighbor surface oxygen to form a second hydroxyl group. The chemisorbed layer is the first formed layer. (b) After chemical completion of the first layer, subsequent water vapor layers are physisorbed on the first-formed hydroxyl layer. After forming the first physisorbed layer, another water molecule adsorbs via double hydrogen bonding to two neighboring hydroxyl groups. This is continued to form multilayer. (c) By forming the more layers, a large amount of water molecules is physisorbed and bonded water vapor molecules become mobile and form continuous dipoles and electrolyte layers between the electrodes to generate dielectric constant and bulk conductivity. **Figure 7** shows the layer by layer adsorption of humidity on the ceramic intelligent surface.

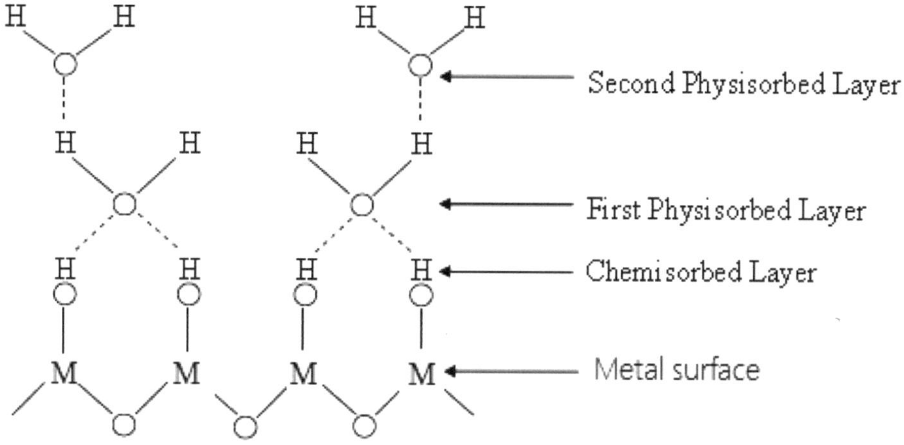

Figure 7. Mechanism of humidity sensing [113].

Ferrite and cobaltate/cobaltite are two main groups of materials used as active sites in CO sensing device. The mechanism of CO sensors is based on an anionic adsorption and the lattice oxygen atoms activities in the surface of perovskites ceramic. Here, the oxygen atoms are adsorbed and react with the available CO gases to form CO_2 and generate a free electron [114] of following types (Eq. 2):

$$\begin{aligned} &O_{ads.} + CO \Rightarrow CO_2 + e \\ &CO_g \Rightarrow CO_{aq.} \Rightarrow CO^{+}_{ads.} + e \\ &CO_{ads.} + O_2 \Rightarrow CO_2 + e \\ &CO_g + O^x_o \Rightarrow CO_2 + e + V^x_o \\ &CO + 1/2\, O_2 + V^x_o + e \Rightarrow CO_2 + O^x_o \end{aligned} \tag{2}$$

A possible mechanism of oxidation of CO by lattice oxygen ions on the surface of La-doped $BaTiO_3$ is shown in **Figure 8.**

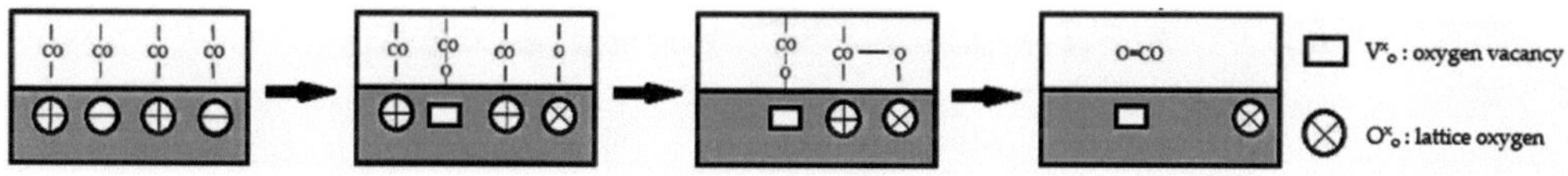

Figure 8. Schematic representation of CO absorption.

LPG (LPG is a mixture of hydrocarbons like n-propane and n-butane), ethanol, and NH_3 sensors are another classes of perovskite-based sensing devices with a variety of applications. The sensing mechanism follows almost the same mechanism. When the perovskite sensor is exposed to air, O_2 adsorbs on the surface substrate and trap electrons from the conduction band of perovskites. This occurs due to the electronegativity of oxygen atom, negative-charged chemisorbed oxygen species, that is, O^{2-} and O^-. As a result, the number of holes increased, the resistance decreased, and the concentration of available carrier achieves a higher value. By exposing the sensing materials to reducing gases, the gas molecules first reacted with the adsorbed oxygen, the carrier (holes) density depressed (due to electron donating nature of gases), and finally increasing the resistance. The reaction mechanisms of LPG (Eq. 3), ethanol (Eq. 4), and NH_3 (Eq. 5) [115] are the following:

$$C_pH_{2p+2} + (3p+1)O^-_{(ads)} \rightarrow pCO_2 + (p+1)H_2O_{(g)} + (3p+1)e^- \quad (3)$$

$$C_2H_5OH + 6O^{n-}_{(ads)} \rightarrow 2CO_2 + 3H_2O + 6ne^- \quad (4)$$

$$2NH_3 + 3O^-_{(ads)} \rightarrow N_2 + 3H_2O + 3e^- \quad (5)$$

An example is $LaCo_xFe_{1-x}O_3$ nanoparticles which used to investigate the ethanol sensing properties [116]. A careful consideration on the response curves indicates that the $LaCo_{0.1}Fe_{0.9}O_3$ nanoparticles are very promising materials for fabricating ethanol sensors (**Figure 9**).

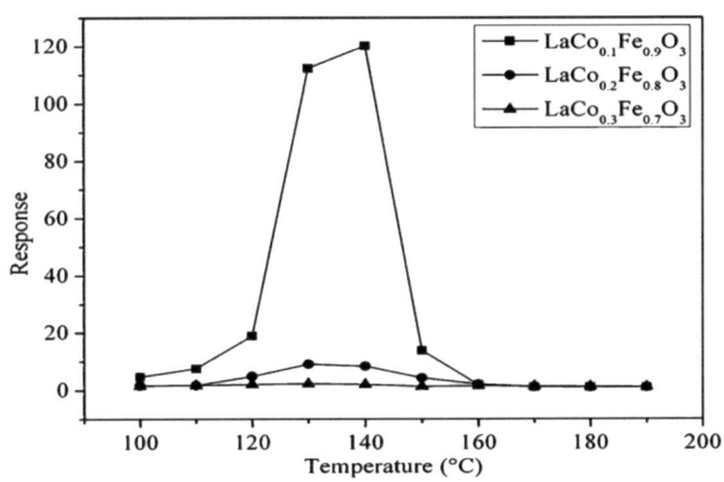

Figure 9. Responses of $LaFeO_3$, $LaCo_{0.1}Fe_{0.9}O_3$, $LaCo_{0.2}Fe_{0.8}O_3$, and $LaCo_{0.3}Fe_{0.7}O_3$ nanoparticles to 500 ppm ethanol [116].

Miscellaneous applications of perovskites in solid-state sensors such as infrared, electro-optic, heat, magnetic field, and liquid-state sensors especially in health-care products such as glucose and cholesterol are reported (**Table 5**). The form of the substrate is thick film including a

http://dx.doi.org/10.5772/62559

nanocomposite of perovskites and a polymer. Typical example of this class is $La_{0.67}Sr_{0.33}MnO_3$ (LSMO). A 400 nm thick LSMO films grow on the lucalox (polycrystalline Al_2O_3 + Mg) substrates *via* vapor deposition technique. Mechanism of such sensors is based on the change of the film resistivity depending on the applied magnetic field magnitude [119]. Initial resistance and the sensitivity to magnetic field are two important factors depend on ambient temperature. This is a drawback for magnetic field measurements because a temperature compensation mechanism is required. **Figure 10** shows a typical magnetic field LSMO sensor. Twisted wires solder to the parts of the electrodes above the substrate (uncovered). Some samples were covered with the polyethylene hot-melt adhesive.

Sensing materials	Infrared	Magnetic field	Heat	Electro-optic	Glucose	H_2O_2	Cholesterol	NADH	Acetone	Refs.
Perovskite										
Titanate		$Ba_{0.25}Sr_{0.75}TiO_3$								[117]
Ferrite					$LaTiO_3$, $ZnFe_2O_4$, $CuFe_2O_4$	$CoFe_2O_4NiFe_2O_4$		$CoFe_2O_4CoFe_2O_4$, $NiFe_2O_4$, $Ni_{1-x}Co_xFe_2O_4$		[47, 98–100, 118–121]
Mange nate	$Ni_xMn_{3-x}O_4$	$La_{0.67}Sr_{0.33}MnO_3$, $La0_{.67}Ba_{0.33}O_3$ LaSrMnO, $La_{2/3}Sr_{1/3}MnO_3$, $La_{0.60}Pb_{0.40}MnO_3$								[119, 122–126]
Niobate			$LiNbO_3$	$KTa_{1-x}NbxO_3$						[45, 127]
Zirconate	$Pb(Zr_{52}Ti_4S)O_3$									[18, 128]

Table 5. Miscellaneous perovskites sensors

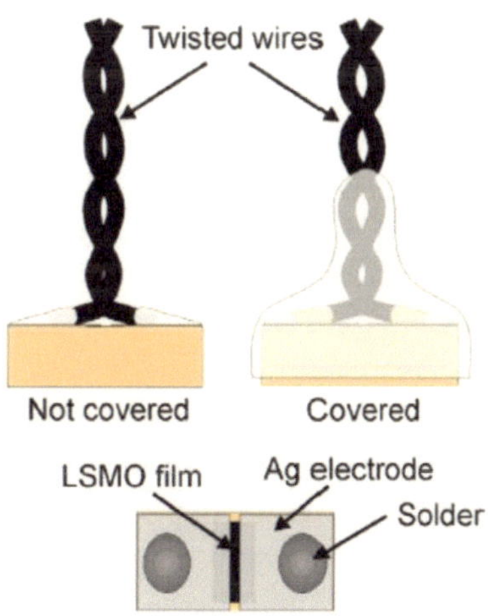

Figure 10. Covered and uncovered LSMO sensor (top and cross-section view) [119].

High selectivity, good sensitivity, fast, and reversible response are the main advantages of enzymatic glucose sensors (EGS) [129]. However, some deficiencies prevented wide range applications of EG sensors like the lack of mechanical and thermal stability and also environmental concerns. Development of glucose sensors without using enzymes based on magnetic nanoparticles can play a significant role. The nanoperovskites can be a good candidate due to the high surface area, high catalytic efficiency, superior mass transport, and strong adsorption ability [120]. Here, an electrode is covered/modified by intelligence perovskite nanomaterials or polymer-perovskite nanocomposite to enhance the sensing properties. In a typical example, polypyrrole is used as shell in polypyrrole–$ZnFe_2O_4$ magnetic nanocomposites to induce a strict barrier and reduce the magnetic coupling effect between nanoparticles (**Figure 11**). Electro-oxidation mechanism of glucose on $ZnFe_2O_4$/PPy core shell electrode is also shown [121].

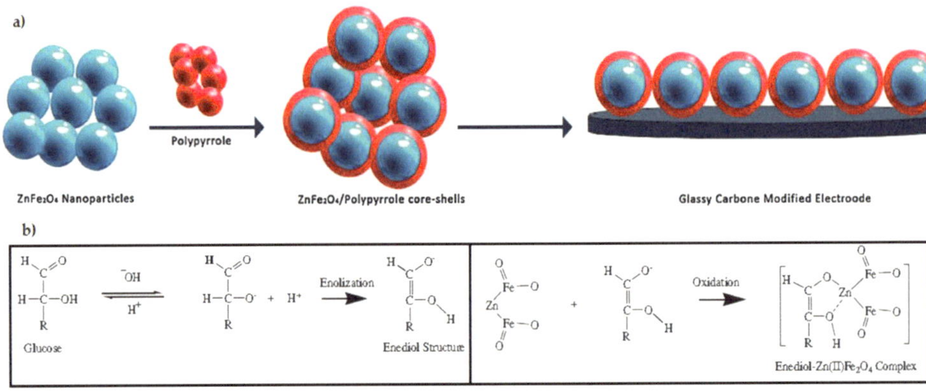

Figure 11 The illustration of (a) preparation process of modified electrode and (b) electro-oxidation mechanism of glucose on $ZnFe_2O_4$/PPy core-shell–glassy carbone modified electrode [121].

5. Fabrication of gas sensor

Various fabrication techniques have been developed in the production of metal-oxide semiconductor sensors. Purity, porosity, reliability, reproducibility, and expense are the most important factors in selecting the production technique [130]. The mechanism of gas sensors are discussed in previous sections. Thick film and thin film technologies have been popular techniques to minify various types of the sensors unifying in monolithic hybrid circuits. In the general definitions, thick film depositions are a process of surface modification by applying a thick coating on a substrate. The manufacture of such devices is multilayer coating film involving deposition of several layers deposition of conductor, resistor, and dielectric layers on the insulating substrate. Thin film technology is a process of deposition of required materials by applying a very thin coating layer often just a few nanometers thickness. Screen printing of ceramic powders, chemical vapor deposition (CVD), sol–gel techniques, spray pyrolysis, physical vapor deposition (PVD), and drop coating [130] are the main techniques for production of metal-oxide films for gas sensors.

http://dx.doi.org/10.5772/62559

5.1. Screen printing

Spiral shape sensing device (screen printing) is one of the simplest sensors in most of the as-reported research articles [13]. Intelligent ink pushes through a porous layer, and it contains the material to be deposited on the substrate. After complete deposition, the print is heated to remove the vehicle, leaving perovskites on the specific target area. **Figure 12** indicates a unit of gas sensor containing metal oxide fabricated by screen printing.

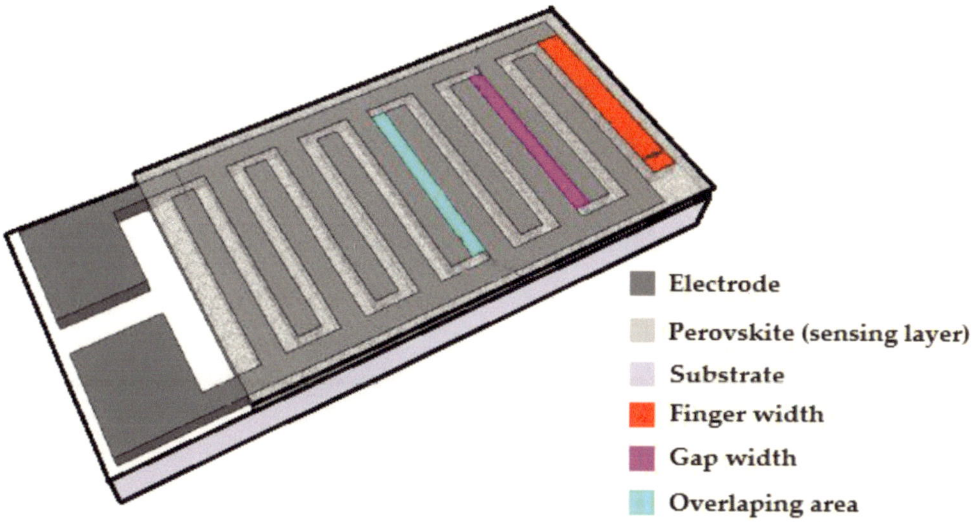

Figure 12. Unit of gas sensor fabricated by screen printing [113].

5.2. Chemical vapor deposition

A heated substrate is exposed to a precursor or controlled mixture of precursors in the vapor phase. The vapors finally interact on the substrate to form a film of the intelligent material. **Figure 13** demonstrates a gas sensors fabricated by CVD.

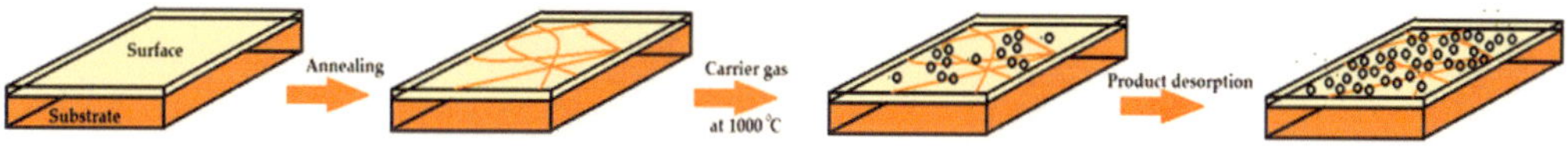

Figure 13. Gas sensor fabricated by chemical vapor deposition process.

5.3. Sol–gel

A film formation by sol–gel involves the formation of colloidal suspension of solid particles and cross-linking between particles, followed by evaporation, and finally heating the film to form a dense surface on the substrate (**Figure 14**).

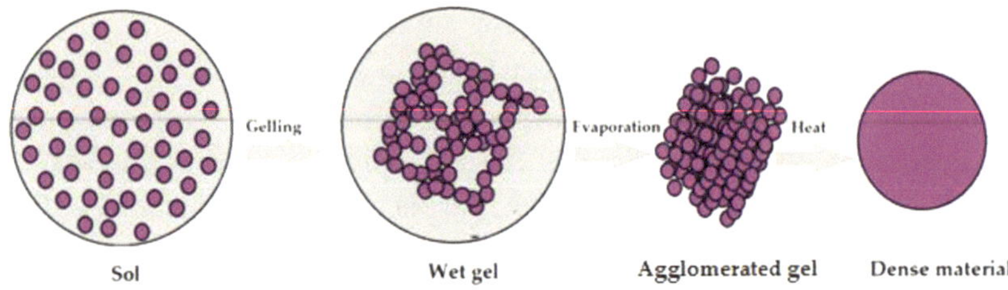

Figure 14. Gas sensing film fabricated via a sol–gel technique.

5.4. Spray pyrolysis

The reactants are sprayed on to the target substrate and then react on the surface by continuous heating to form film (**Figure 15**).

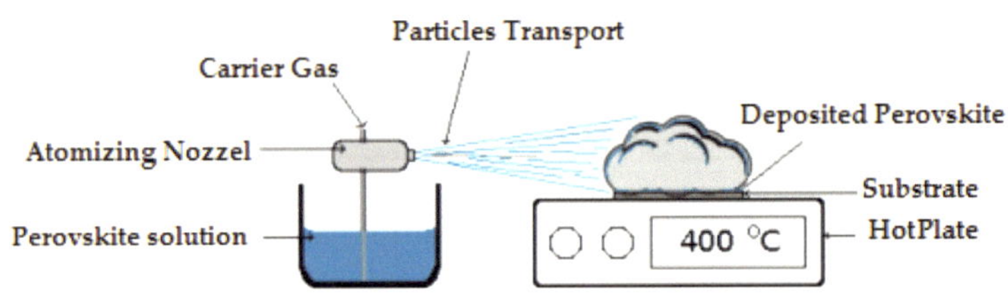

Figure 15. Schematic representation of spray pyrolysis [131].

5.5. Physical vapor deposition

The starting materials are transferred into the gas phase by evaporation/sputtering. A reactive gas reacts with the metal vapor, to form a compound, which further deposited on the substrate (**Figure 16**).

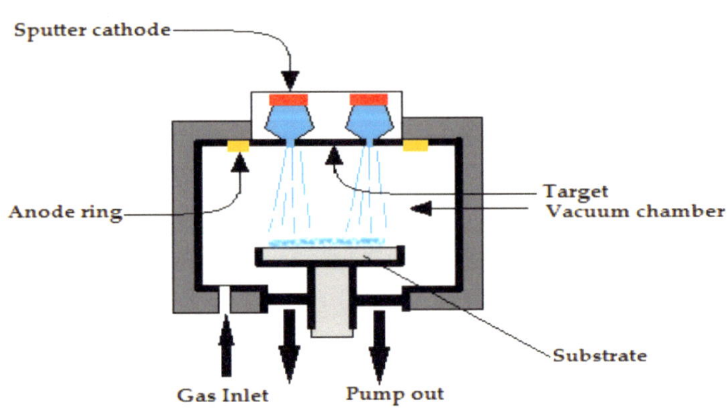

Figure 16. Unit of physical vapor deposition.

5.6. Drop coating

Drop coating is a process by where a paste is made of the desired perovskite powder and a solvent; the mixture is then deposited onto a substrate using an appropriate apparatus followed by the evaporation to form a film (**Figure 17**).

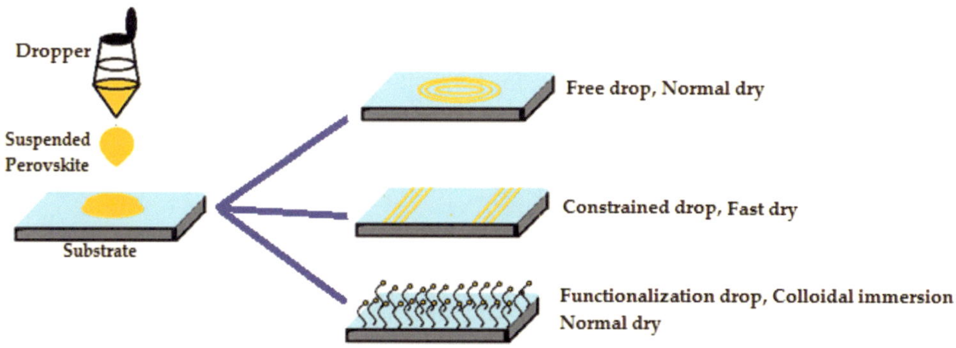

Figure 17. Drop coating sensor fabrication.

How does a real gas-sensor response to the analyte? The gas-sensor performance measures from the V_{out} of R_L that cascades R_s (the resistance of gas sensor). The resistance of gas sensor (R_s) and the sensor response (S) obtain from the (Eqs. 6 and 7), respectively;

$$R_S = R_L \left(V_c - V_{out}\right) / V_{out.} \tag{6}$$

$$S = R_a / R_g \tag{7}$$

where R_a is the resistance in air and R_g is the resistance in the air mixed with detected gases. The response time is the time required for the conductance to reach 90% of the equilibrium value after a test gas injection, and the recovery time was the time necessary for a sensor to attain a conductance 10% above its original value in air (**Figure 18**).

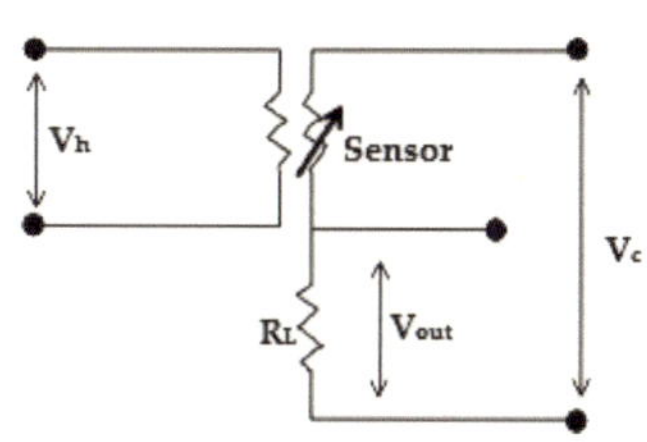

Figure 18. Electrical circuit diagram of a gas sensor [132].

For instance, $YCoO_3$-based sensors (p-type semiconductor) were tested at temperature range of 100–380°C. The obtained response to the gas (here NO_2) versus time can be given by the contribution of two different reactions: (a) oxidation and (b) reducing of the surface (**Figure 19**). The former is faster and favored at low temperature, and the latter is slower and favored at higher temperatures [133].

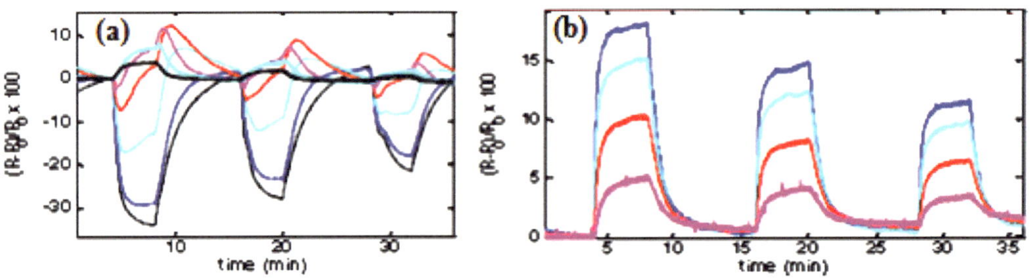

Figure 19. Typical response to (a) NO_2 and (b) CO at various temperature [133].

Artificial olfactory systems (AOS) based on metal-oxides chemiresistors is one of the devices which practically used for sensing the air pollutions. $YCoO_3$ sensors doped with various metals were selected for CO and NO_2 detection at different temperatures. At around 280°C, the sensor-doped Pd shows a satisfactory sensitivity with a large response speed to CO gas. The same electrode-doped Co was selected for NO_2 pollution sensing. The results clearly indicate that a good sensitivity and a fast response (response time, $T_{10\text{–}90\%}$ = 1 min, and recovery time, $T_{90\text{–}10\%}$ = 3 min) at a temperature close to 180 °C [133].

The selectivity toward various target gases such as sulfur dioxide (SO_2), ammonia (NH_3), hydrogen sulfide (H_2S), nitrogen dioxide (NO_2) has been discussed before. For instance, 95% sensing response of $CoFe_2O_4$ observes toward NO_2 compared to other gases at an operating temperature of 150°C, 5 seconds response time and 117 seconds recovery time with NO_2 gas [23]. Performing the same sensors for SO_2, NH_3, and H_2S was given just 3%, 6%, and 7%, respectively. The ferrites such as $NiFe_2O_4$, $ZnFe_2O_4$, $MgFe_2O_4$, $ZnAl_2O_4$, $CoAl_2O_4$, and $MgAl_2O_4$ shows the gas response 10–20% for H_2S and 1–10% for NH_3 gas at 300°C operating temperature.

The sensing response of $GaFeO_3$ toward ethanol vapor at 350°C demonstrates that the resistance declines rapidly in the alcohol atmosphere which exhibited n-type behavior. The response characteristics of the sensors below 350°C for 1 ppm of methanol, ethanol, and isopropanol do not show any significant output. This is due to low thermal energy of the molecules to react with the surface adsorbed oxygen species [105]. **Table 6** illustrates some typical experimental results of gas sensing properties of perovskite-based materials.

Perovskite	Sensing material	Sensing response (%)	Response time (Sec.)	Recovery time (Sec.)	Response temperature (°C)	Refs.
$LaCoO_3$	CO	95	423	377	250	[134]
$AlNbO_3$	Ethanol	99	24	100	250	[109]
$ZnSnO_3$	H_2	95	1	12	>375	[135]
$ZnWO_4$	H_2O	98	15	158	R.T	[90]
$NiFe_2O_4$	H_2S	75	60	300	150	[136]
$CoWO_4$	NH_3	90	3	1	Elevated	[118]
$SmFe_{1-x}Co_xO_3$	O_3	<80	–	600	R.T	[137]
$La_{1-x}Sr_xFeO_3$	CO_2	<80	660	300	380	[138]
$LaFeO_3$	C_4H_{10}	93	–	–	250	[139]
$CuFe_2O_4$	LPG	86	5	68	350	[11]
Zn_2SnO_4	NO_2	–	43	326	200	[79]

Table 6. Practical efficiencies of some important perovskites-based sensors.

6. Summary and recommendation for future research

The oxygen partial pressure dependence of the point defect concentration, and its conductivity, in perovskite semiconductors allows for their application in sensors. The resistance of the as-mentioned semiconductors can be affected by other gases, such as O_2, CO, NH_3, hydrocarbons, and ethanol, which create opportunities for developing new chemical sensors.

In addressing diversity and establishing the current chapter, it is necessary to develop the sensors technology by routine reliable synthesis of self-assembled materials including inorganic materials or inorganic–organic nanocomposites for real-life applications and also commercialization. The need of novel perovskites sensing materials with suitable compositions from all reported perovskites (summarized in **Tables 4** and **5**) as a filler/cover for sensing device for an especial chemical. The need of high-accuracy sensing device in everyday life especially for health care and also environmental concerns requires a systematic management and researches. For example, a suitable combination of perovskite materials can be identified out of 15 components for ethanol sensors. Moreover, suitable manufacturing technologies, free choice of device geometrical properties to attain the indispensable dimensional efficiencies, optimization of surface for the occurrence of conductance, ease of production flow, and investment expenses are immediately required. It seems that the accurate optimization of perovskite materials applicable in gas sensor is influenced by temperature and partial pressure of oxygen. Therefore, the performance of gas sensor depends on the exact height from the sea level.

Author details

Morteza Enhessari* and Ali Salehabadi

*Address all correspondence to: enhessari@gmail.com/enhessari@iau-naragh.ac.ir

Department of Chemistry, Naragh Branch, Islamic Azad University, Naragh, Iran

References

[1] Atkins P., Overton T., Rourke J., Weller M., Armstrong F., Inorganic Chemistry, Fifth Edition, Oxford University Press, USA, 2006.

[2] Qi T., Levchenko SV, Bennett JW, Grinberg I., Rappe AM, New Prospects for High Performance SONAR, Chemical Sensor, and Communication Device Materials, DoD High Performance Computing Modernization Program Users Group, IEEE (2010) 197–204.

[3] Shimizu Y., Uemura K., Matsuda H., Miura N., Yamazoe N., Bi-functional oxygen electrode using large surface area $La_{1-x}Ca_xCoO_3$ for rechargeable metal-air battery, J. Electrochem. Soc. 137 (1990) 3430–3433.

[4] Hunter GW, Xu JC, Evans LJ, Vander Wal RL, Gordon M., Chemical sensors based on metal oxide nanostructures, applications: Solid-state ionic devices, 9(2006)199–209.

[5] Shaterian M., Enhessari M., Rabbani D., Asghari M., Synthesis, characterization and photocatalytic activity of $LaMnO_3$ nanoparticles, Appl. Surf. Sci. 318 (2014) 213–217.

[6] Ghasdi M., Alamdari H., CO sensitive nanocrystalline $LaCoO_3$ perovskite sensor prepared by high energy ball milling, Sensors Actuators B Chem. 148 (2010) 478–485.

[7] Wang L., Kumar RV, Thick film miniaturized HCl gas sensor, Sensors Actuators B Chem. 98 (2004) 196–203.

[8] Pokhrel S., Huo L., Zhao H., Gao S., Thick film of $LaCr_{1-x}Ti_xO_3$ (x = 0.4) perovskites prepared by combustion technique for alcohol sensing application, Sensors Actuators B Chem. 122 (2007) 321–327.

[9] Luyten J., Chemical and electrical properties of Yb-doped strontium cerates in coal combustion atmospheres, Solid State Ion. 46 (1991) 117–120.

[10] Guoxi X., Yuebin X., Effects on magnetic properties of different metal ions substitution cobalt ferrites synthesis by sol–gel auto-combustion route using used batteries, Mater. Lett. 164 (2016) 444–448.

[11] Khandekar MS, Tarwal NL, Patil JY, Shaikh FI, Mulla IS, Suryavanshi SS, Liquefied petroleum gas sensing performance of cerium doped copper ferrite, Ceram. Int. 39 (2013) 5901–5907.

[12] Enhessari M., Khanahmadzadeh S., Ozaee K., Structural characterization of $BaZrO_3$ nanopowders prepared by stearic acid gel method, J. Iran. Chem. Res. 3 (2010) 11–15.

[13] Shandiz RH, Shaterian M., Ozaee K., Enhessari M., Fabrication of low temperature gas sensor using $SrCe_{0.9}Yb_{0.1}O_3$ nanopowders as proton conductor, Synth. React. Inorg. Met. Nano-Metal Chem. 45 (2013) 1108–1111.

[14] Shaterian M., Barati M., Ozaee K., Enhessari M., Application of $MnTiO_3$ nanoparticles as coating layer of high performance $TiO_2/MnTiO_3$ dye-sensitized solar cell, J. Ind. Eng. Chem. 20 (2013) 3646–3648.

[15] Enhessari M., Salehabadi A., Nasrollahi Z., Ozaee K., Ammonia-assisted synthesis method for $CoTiO_3$ nanoporous matrix, J. Semicond. 37 (2016) 1–3.

[16] Huang X., Choi Y., Chemical sensors based on nanostructured materials, Sensors and Actuators B, 122 (2006) 659–671.

[17] Mukherjee K., Majumder SB, Hydrogen sensing characteristics of nano-crystalline $Mg_{0.5}Zn_{0.5}Fe_2O_4$ thin film: Effect of film thickness and operating temperature, Int. J. Hydrogen Energy 39 (2014) 1185–1191.

[18] Chang CC, Tang CS, An integrated pyroelectric infrared sensor with a PZT thin film, Sensors Actuators A Phys. 65 (1998) 171–174.

[19] Shi C., Qin H., Zhao M., Wang X., Li L., Hu J., Investigation on electrical transport, CO sensing characteristics and mechanism for nanocrystalline $La_{1-x}Ca_xFeO_3$ sensors, Sensors Actuators B Chem. 190 (2014) 25–31.

[20] Andre RS, Zanetti SM, Varela JA, Longo E., Synthesis by a chemical method and characterization of $CaZrO_3$ powders: Potential application as humidity sensors, B2 Ceram. Int. 40 (2014) 16627–16634.

[21] Chow CL, Huang H., Ang WC, Liu H., Huang Y., Tse MS, et al., Effect of annealing temperature on the crystallization and oxygen sensing property of strontium titanate ferrite sol–gel thin films, Sensors Actuators B Chem. 187 (2013) 20–26.

[22] Roffat M., Noël O., Soppera O., Bohnke O., Investigation of the perovskite ceramic $Li_{0.30}La_{0.56}Ti\ O_3$ by Pulsed Force Mode AFM for p H sensor application, Sensors Actuators B Chem. 138 (2009) 193–200.

[23] Bagade AA., Rajpure KY, Development of $CoFe_2O_4$ thin films for nitrogen dioxide sensing at moderate operating temperature, J. Alloys Compd. 657 (2016) 414–421.

[24] Dharmadhikari DV, Nikam SK, Athawale AA, Template free hydrothermal synthesis and gas sensing application of lanthanum cuprate (La_2CuO_4): Effect of precursors on phase formation and morphology, J. Alloys Compd. 590 (2014) 486–493.

[25] Enhessari M., Razi MK, Etemad L., Parviz A., Sakhaei M., Structural, optical and magnetic properties of the Fe_2TiO_5 nanopowders, J. Exp. Nanosci. 9 (2014) 167–176.

[26] Sadjadi MS, Mozaffari M., Enhessari M., Zare K., Superlattices and microstructures effects of $NiTiO_3$ nanoparticles supported by mesoporous MCM-41 on photoreduction of methylene blue under UV and visible light irradiation, Superlattices Microstruct. 47 (2010) 685–694.

[27] Sadjadi M.S., Zare K., Khanahmadzadeh S., Enhessari M., Structural characterization of $NiTiO_3$ nanopowders prepared by stearic acid gel method, Mater. Lett. 62 (2008) 3679–3681.

[28] Enhessari M., Parviz A., Ozaee K., Karamali E., Magnetic properties and heat capacity of $CoTiO_3$ nanopowders prepared by stearic acid gel method, J. Exp. Nanosci. 5 (2010) 61–68.

[29] Enhessari M., Shaterian M., Esfahani MJ, Motaharian MN, Synthesis, characterization and optical band gap of La_2CuO_4 nanoparticles, Mater. Sci. Semicond. Process. 16 (2013) 1517–1520.

[30] Zare K., Sadjadi MS, Enhessari M., Khanahmadzadeh S., Synthesis and characterization of $PbTiO_3$ nanopowders by citric acid gel method, J. Phys. Theor. Chem. 6 (2009) 9–12.

[31] Ghasdi M., Alamdari H., Royer S., Adnot A., Electrical and CO gas sensing properties of nanostructured $La_{1-x}Ce_xCoO_3$ perovskite prepared by activated reactive synthesis, Sensors Actuators B Chem. 156 (2011) 147–155.

[32] Zhou L., Li X., Wu H., Liao Z., Yuan Q., Xia F., et al., Sensing properties of YSZ-based NOx sensors with double-perovskite $(La_{0.8}Sr_{0.2})_2FeNiO_6$-sensing electrodes, Ceram. Int. 40 (2014) 9257–9263.

[33] Mulmi S., Kannan R., Thangadurai V., CO_2 and SO_2 tolerant Fe-doped metal oxides for solid state gas sensors, Solid State Ion. 262 (2014) 274–278.

[34] Lancaster D., Weidner R., Richter D., Tittel F., Limpert J., Compact CH4 sensor based on difference frequency mixing of diode lasers in quasi-phasematched $LiNbO_3$, Opt. Commun. 175 (2000) 461–468.

[35] Fergus JW, Perovskite oxides for semiconductor-based gas sensors, Sensors Actuators B Chem. 123 (2007) 1169–1179.

[36] Wang X., Qin H., Sun L., Hu J., CO_2 sensing properties and mechanism of nanocrystalline $LaFeO_3$ sensor, Sensors Actuators B Chem. 188 (2013) 965–971.

[37] Patil LA, Suryawanshi DN, Pathan IG, Patil DG, Effect of firing temperature on gas sensing properties of nanocrystalline perovskite $BaTiO_3$ thin films prepared by spray pyrolysis techniques, Sensors Actuators B Chem. 195 (2014) 643–650.

[38] Giang HT, Duy HT, Ngan PQ, Thai GH, Thi D., Thu A., High sensitivity and selectivity of mixed potential sensor based on Pt/YSZ/$SmFeO_3$ to NO_2 gas, Sensors Actuators B Chem. 183 (2013) 550–555.

[39] Giang HT, Duy HT, Ngan PQ, Thai GH, Thi D., Thu A., et al., Hydrocarbon gas sensing of nano-crystalline perovskite oxides $LnFeO_3$ (Ln = La, Nd and Sm), Sensors Actuators B Chem. 158 (2011) 246–251.

[40] Dhivya P., Prasad AK, Sridharan M., Nanostructured perovskite $CdTiO_3$ films for methane sensing, Sensors Actuators B. Chem. 222 (2015) 987–993.

[41] Ishihara T., Perovskite Oxide for Solid Oxide Fuel Cells, Springer, New York, USA, 2009.

[42] Armstrong EN, Striker T., Ramaswamy V., Ruud JA, Wachsman ED, NO_x adsorption behavior of $LaFeO_3$ and $LaMnO_{3+\delta}$ and its influence on potentiometric sensor response, Sensors Actuators B Chem. 158 (2011) 159–170.

[43] Michael L., Brian W., Post ML, Sanders BW, Kennepohl P., Thin films of non-stoichiometric perovskites as potential oxygen sensors, Sensors Actuators B Chem. 1 (1993) 272–275.

[44] Post M., Tunney J., Yang D., Du X., Singleton D., Material chemistry of perovskite compounds as chemical sensors, Sensors Actuators B Chem. 59 (1999) 190–194.

[45] Sarker MRH, Karim H., Martinez R., Delfin D., Enriquez R., Shuvo MAI, Temperature measurements using a lithium niobate ($LiNbO_3$) pyroelectric ceramic, Measurement 75 (2015) 104–110.

[46] Tasaki T., Takase S., Shimizu Y., Impedance metric acetylene gas sensing properties of Sm–Fe-based perovskite-type oxide-based thick-film device, Sensors Actuators B Chem. 187 (2013) 128–134.

[47] Gupta R., YBCO-FET room temperature ammonia sensor, Sensors Actuators B Chem. 63 (2000) 35–41.

[48] Deng J., Zhu W., Tan OK, Yao X., Amorphous Pb(Zr,Ti)O_3 thin film hydrogen gas sensor, Sensors and Actuators B, 77 (2001) 416–420.

[49] Yannopoulos LN, A p-type semiconductor thick film gas sensor, Sensors Actuators 12 (1987) 263–273.

[50] Šutka A., Gross KA, Spinel ferrite oxide semiconductor gas sensors, Sensors Actuators B Chem. 222 (2016) 95–105.

[51] Choopun S., Wongrat E., N. Hongsith, Metal-oxide nanowires for gas sensors, Met-Oxide Nanowires Gas Sensors, 2012: pp. 3–24.

[52] Argirusis C., Jomard F., Wagner SF, Menesklou W., Ivers-Tiffée E., Study of the oxygen incorporation and diffusion in $Sr(Ti_{0.65}Fe_{0.35})O_3$ ceramics, Solid State Ion. 192 (2011) 9–11.

[53] Isarakorn D., Briand D., Sambri A., Gariglio S., Triscone JM, Guy F., et al., Finite element analysis and experiments on a silicon membrane actuated by an epitaxial PZT thin film for localized-mass sensing applications, Sensors Actuators B Chem. 153 (2011) 54–63.

[54] Wang L., He Y., Hu J., Qi Q., Zhang T., DC humidity sensing properties of $BaTiO_3$ nanofiber sensors with different electrode materials, Sensors Actuators B Chem. 153 (2011) 460–464.

[55] Imran Z., Batool SS, Jamil H., Usman M., Israr-Qadir M., Shah SH, et al., Excellent humidity sensing properties of cadmium titanate nanofibers, Ceram. Int. 39 (2013) 457–462.

[56] Zhang Y., Fu W., Yang H., Li M., Li Y., Zhao W., et al., A novel humidity sensor based on $Na_2Ti_3O_7$ nanowires with rapid response-recovery, Sensors Actuators B Chem. 135 (2008) 317–321.

[57] J. Yoon, G. Hunter, S. Akbar, Dutta PK, Interface reaction and its effect on the performance of a CO_2 gas sensor based on $Li_{0.35}La_{0.55}Ti\ O_3$ electrolyte and Li_2CO_3 sensing electrode, Sensors Actuators B Chem. 182 (2013) 95–103.

[58] Chapelle A., Yaacob MH, Pasquet I., Presmanes L., BarnabéA., Tailhades P., et al., Structural and gas-sensing properties of $CuO–Cu_xFe_{3-x}O_4$ nanostructured thin films, Sensors Actuators B Chem. 153 (2011) 117–124.

[59] Mukherjee K., Majumder SB, Synthesis of embedded and isolated $Mg_{0.5}Zn_{0.5}Fe_2O_4$ nanotubes and investigation on their anomalous gas sensing characteristics, Sensors Actuators B Chem. 177 (2013) 55–63.

[60] Meuffels P., Propane gas sensing with high-density $SrTi_{0.6}Fe_{0.4}O_{(3-\delta)}$ ceramics evaluated by thermogravimetric analysis, J. Eur. Ceram. Soc. 27 (2007) 285–290.

[61] Iio A., Ikeda H., Anggraini SA, Miura N., Potentiometric YSZ-based oxygen sensor using $BaFeO_3$ sensing-electrode, Electrochem. Commun. 48 (2014) 134–137.

[62] Cavalieri A., Caronna T., Natali Sora I., Tulliani JM, Electrical characterization of room temperature humidity sensors in $La_{0.8}Sr_{0.2}Fe_{1-x}Cu_xO_3$ (x =0, 0.05, 0.10), Ceram. Int. 38 (2012) 2865–2872.

[63] Casas-Cabanas M., Chadwick AV, Palacín MR, Michel CR, Carbon dioxide sensing properties of bismuth cobaltite, Sensors Actuators B Chem. 157 (2011) 380–387.

[64] Gómez L., Galeano V., Parra R., Michel C.R., Paucar C., Morán O., Carbon dioxide gas sensing properties of ordered oxygen deficient perovskite $LnBaCo_2O_{5+\delta}$ (Ln = La, Eu), Sensors Actuators B Chem. 221 (2015) 1455–1460.

[65] Malavasi L., Tealdi C., Flor G., Chiodelli G., Cervetto V., Montenero A., et al., $NdCoO_3$ perovskite as possible candidate for CO-sensors: Thin films synthesis and sensing properties, Sensors Actuators B Chem. 105 (2005) 407–411.

[66] Michel CR, Delgado E., Martínez AH, Evidence of improvement in gas sensing properties of nanostructured bismuth cobaltite prepared by solution-polymerization method, Sensors Actuators B Chem. 125 (2007) 389–395.

[67] Shuk P., Vecher A., Kharton V., Tichonova L., Wiemhöfer HD, Guth U., et al., Electrodes for oxygen sensors based on rate earth manganites or cabaltites, Sensors Actuators B Chem. 16 (1993) 401–405.

[68] Tealdi C., Islam MS, Fisher CAJ., Malavasi L., Flor G., Defect and transport properties of the $NdCoO_3$ catalyst and sensor material, Prog. Solid State Chem. 35 (2007) 491–499.

[69] Wei X., Wei T., Li J., Lan X., Xiao H., Lin YS, Strontium cobaltite coated optical sensors for high temperature carbon dioxide detection, Sensors Actuators B Chem. 144 (2010) 260–266.

[70] Mullen MR, Spirig JV, Hoy J., Routbort JL, Singh D., Dutta PK, Development of nanosized lanthanum strontium aluminum manganite as electrodes for potentiometric oxygen sensor, Sensors Actuators B Chem. 203 (2014) 670–676.

[71] Wei XT, Wei T., Xiao H., Lin YS, Terbium doped strontium cerate enabled long period fiber gratings for high temperature sensing of hydrogen, Sensors Actuators B Chem. 152 (2011) 214–219.

[72] Schutter FD, Vangrunderbeek J., Luyten J., Proton conductivity in strontium cerates for hydrogen gas sensors in coal gasification systems, Solid State Ionics. 57 (1992) 77–81.

[73] Zhou H., Dai L., Jia L., Zhu J., Li Y., Wang L., Effect of fluorine, chlorine and bromine doping on the properties of gadolinium doped barium cerate electrolytes, Int. J. Hydrogen Energy 40 (2015) 8980–8988.

[74] Zhang M., Hu C., Liu H., Xiong Y., Zhang Z., A rapid-response humidity sensor based on $BaNbO_3$ nanocrystals, Sensors Actuators B Chem. 136 (2009) 128–132.

[75] Gnanasekar KI, Jayaraman V., Prabhu E., Gnanasekaran T., Periaswami G., Electrical and sensor properties of $FeNbO_4$: A new sensor material, Sensors Actuators B Chem. 55 (1999) 170–174.

[76] Xuchen L., Preparation and characterization of $LaNiO_3$ A r F ratio-sensitive thin film by sol–gel process based on amorphous citrate precursors, (2000) 24–28.

[77] Kocemba I., Wróbel-Jedrzejewska M., Szychowska A., Rynkowski J., Główka M., The properties of barium stannate and aluminum oxide-based gas sensor. The role of Al_2O_3 in this system, Sensors Actuators B Chem. 121 (2007) 401–405.

[78] Song P., Wang Q., Yang Z., Ammonia gas sensor based on PPy/$ZnSnO_3$ nanocomposites, Mater. Lett. 65 (2011) 430–432.

[79] Ganbavle VV, Patil MA, Deshmukh HP, Rajpure KY, Development of $Zn_2Sn\ O_4$ thin films deposited by spray pyrolysis method and their utility for NO_2 gas sensors at moderate operating temperature, J. Anal. Appl. Pyrolysis 107 (2014) 233–241.

[80] Huang J., Xu X., Gu C., Wang W., Geng B., Sun Y., et al., Size-controlled synthesis of porous $ZnSnO_3$ cubes and their gas-sensing and photocatalysis properties, Sensors Actuators B Chem. 171–172 (2012) 572–579.

[81] Jiang YQ, Chen XX, Sun R., Xiong Z., Zheng LS, Hydrothermal syntheses and gas sensing properties of cubic and quasi-cubic $Zn_2Sn\ O_4$, Mater. Chem. Phys. 129 (2011) 53–61.

[82] Tharsika T., Haseeb ASMA, Akbar SA, Sabri MFM, Wong YH, Gas sensing properties of zinc stannate ($Zn_2Sn\ O_4$) nanowires prepared by carbon assisted thermal evaporation process, J. Alloys Compd. 618 (2015) 455–462.

[83] Upadhyay S., Kavitha P., Lanthanum doped barium stannate for humidity sensor, Mater. Lett. 61 (2007) 1912–1915.

[84] Zhou M., Ahmad A., Sol–gel processing of In-doped $CaZrO_3$ solid electrolyte and the impedimetric sensing characteristics of humidity and hydrogen, Sensors Actuators B Chem. 129 (2008) 285–291.

[85] Saha D., Giri R., Mistry KK, Sengupta K., Magnesium chromate-TiO_2 spinel tape cast thick film as humidity sensor, Sensors Actuators B Chem. 107 (2005) 323–331.

[86] Pingale S., Patil S., Vinod M., Pathak G., Vijayamohanan K., Mechanism of humidity sensing of Ti-doped $MgCrO_3$ ceramics, Materials Chemistry and Physics. 46 (1996) 72–76.

[87] Sears WM, The gas-sensing properties of sintered bismuth iron molybdate catalyst, Sensors Actuators 19 (1989) 351–370.

[88] Edwin Suresh Raj AM, Mallika C., Swaminathan K., Sreedharan OM, Nagaraja KS, Zinc(II) oxide-zinc(II) molybdate composite humidity sensor, Sensors Actuators B Chem. 81 (2002) 229–236.

[89] Sundaram R., Nagaraja KS, Solid state electrical conductivity and humidity sensing studies on metal molybdate–molybdenum trioxide composites (M = Ni^{2+}, Cu^{2+} and Pb^{2+}), Sensors Actuators B Chem. 101 (2004) 353–360.

[90] You L., Cao Y., Sun YF, Sun P., Zhang T., Du Y., et al., Humidity sensing properties of nanocrystalline $ZnWO_4$ with porous structures, Sensors Actuators B Chem. 161 (2012) 799–804.

[91] Gonzalez CM, Du X., Dunford JL, Post ML, Copper tungstate thin-films for nitric oxide sensing, Sensors Actuators B Chem. 173 (2012) 169–176.

[92] Tamaki J., Fujii T., Fujimori K., Miura N., Yamazoe N., Application of metal tungstate-carbonate composite to nitrogen oxides sensor operative at elevated temperature, Sensors Actuators B Chem. 25 (1995) 396–399.

[93] Solis V., Lantto JL, A study of gas-sensing properties of sputtered films, Sensors Actuators B 25 (1995) 591–595.

[94] Solis JL, Lantto V., Haggstrom L., Kalska B., Frantti J., Saukko S., Synthesis of new compound semiconductors in the Sn–W–O system for gas-sensing studies, Sensors Actuators B 68 (2000) 286–292.

[95] Kärkkänen I., Kodu M., Avarmaa T., Kozlova J., Matisen L., Mändar H., et al., Sensitivity of $CoWO_4$ thin films to CO, Procedia Eng. 5 (2010) 160–163.

[96] Suresh Raj AME, Mallika C., Sreedharan O., Nagaraja K., Manganese oxide–manganese tungstate composite humidity sensors, Mater. Lett. 53 (2002) 316–320.

[97] Yadav BC, Yadav A., Singh S., Singh K., Nanocrystalline zinc titanate synthesized via physicochemical route and its application as liquefied petroleum gas sensor, Sensors Actuators B Chem. 177 (2013) 605–611.

[98] Santhaveesuk T., Gardchareon A., Wongratanaphisan D., Choopun S., Ethanol sensing properties of $Zn_2Ti\ O_4$ particles, Ceram. Int. 41 (2015) S809–S813.

[99] Du J., Wu Y., Choy K-L, Shipway PH, Structure, properties and gas sensing behavior of $Cr_{2-x}Ti_xO_3$ films fabricated by electrostatic spray assisted vapour deposition, Thin Solid Films 519 (2010) 1293–1299.

[100] Belle CJ, Wesch GE, Neumeier S., Lozano-Rodríguez MJ, Scheinost AC, Simon U., Volume-doped cobalt titanates for ethanol sensing: An impedance and X-ray absorption spectroscopy study, Sensors Actuators B Chem. 192 (2014) 60–69.

[101] Sutka A., Mezinskis G., Lusis A., Stingaciu M., Gas sensing properties of Zn-doped p-type nickel ferrite, Sensors Actuators B Chem. 171–172 (2012) 354–360.

[102] Franke D., Zosel J., Guth U., NO sensitivity of perovskite-type electrode materials $La_{0.6}Ca_{0.4}B'_{1-x}B''_xO_{3\pm\delta}$ (B′ = Mn, Cr; B″ = Ni, Fe, Co; x = 0, 0.1, …, 0.6) in mixed potential sensors, Sensors Actuators B Chem. 223 (2016) 723–729.

[103] Lee I., Jung B., Park J., Lee C., Hwang J., Park CO, Mixed potential NH_3 sensor with $LaCoO_3$ reference electrode, Sensors Actuators B Chem. 176 (2013) 966–970.

[104] Khandekar MS, Tarwal NL, Mulla IS, Suryavanshi SS, Nanocrystalline Ce doped $CoFe_2O_4$ as an acetone gas sensor, Ceram. Int. 40 (2014) 447–452.

[105] Sen S., Chakraborty N., Rana P., Narjinary M., Mursalin SD, Tripathy S., et al., Nanocrystalline gallium ferrite: A novel material for sensing very low concentration of alcohol vapour, Ceram. Int. 41 (2015) 10110–10115.

[106] Kazin AP, Rumyantseva MN, Prusakov VE, Suzdalev IP, Gaskov AM, Nanocrystalline ferrites $Ni_xZn_{1-x}Fe_2O_4$: Influence of cation distribution on acidic and gas sensing properties, J. Solid State Chem. 184 (2011) 2799–2805.

[107] Balamurugan C., Vijayakumar E., Subramania A., Synthesis and characterization of In Nb O_4 nanopowder for gas sensors, Talanta 88 (2012) 115–20.

[108] Zhu Q., Shih W-H, Shih WY, Enhanced dimethyl methylphosphonate (DMMP) detection sensitivity by lead magnesium niobate-lead titanate/copper piezoelectric microcantilever sensors via Young's modulus change, Sensors Actuators B Chem. 182 (2013) 147–155.

[109] Balamurugan C., Subashini A., Chaudhari GN, Subramania A., Development of wide band gap sensor based on $AlNbO_4$ nanopowder for ethanol, J. Alloys Compd. 526 (2012) 110–115.

[110] Zhang D., Zhang R., Xu C., Fan Y., Yuan B., Microwave-assisted solvothermal synthesis of nickel molybdate nanosheets as a potential catalytic platform for NADH and ethanol sensing, Sensors Actuators B Chem. 206 (2015) 1–7.

[111] Raju AR, Rae CNR, MoO_3/TiO_2 and Bi_2MoO_6 as ammonia sensors, Sensors Actuators B Chem. 21 (1994) 23–26.

[112] Sivapunniyam A., Wiromrat N., Myint MTZ, Dutta J., High-performance liquefied petroleum gas sensing based on nanostructures of zinc oxide and zinc stannate, Sensors Actuators B Chem. 157 (2011) 232–239.

[113] Farahani H., Wagiran R., Hamidon MN, Humidity sensors principle, mechanism, and fabrication technologies: A comprehensive review, Sensors (2014) 7881–7939.

[114] Zhou ZG, Tang ZL, Zhang ZT, Wlodarski W., Perovskite oxide of PTCR ceramics as chemical sensors, Sensors Actuators B Chem. 77 (2001) 22–26.

[115] Balamurugan C., Bhuvanalogini G., Subramania A., Development of nanocrystalline $CrNbO_4$ based p-type semiconducting gas sensor for LPG, ethanol and ammonia, Sensors Actuators B Chem. 168 (2012) 165–171.

[116] Feng C., Ruan S., Li J., Zou B., Luo J., Chen W., et al., Ethanol sensing properties of $LaCo_xFe_{1-x}O_3$ nanoparticles: Effects of calcination temperature, Co-doping, and carbon nanotube-treatment, Sensors Actuators B. Chem. 155 (1991) 232–238.

[117] Huck C., Poghossian A., Bäcker M., Chaudhuri S., Zander W., Schubert J., et al., Capacitively coupled electrolyte-conductivity sensor based on high-k material of barium strontium titanate, Sensors Actuators B Chem. 198 (2014) 102–109.

[118] Diao Q., Yang F., Yin C., Li J., Yang S., Liang X., et al., Ammonia sensors based on stabilized zirconia and $CoWO_4$ sensing electrode, Solid State Ion. 225 (2012) 328–331.

[119] StankevičV., ŠimkevičiusČ., Keršulis S., Balevičius S., ŽurauskienėN., Pavilonis D., et al., Improvement in the long-term stability of parameters of encapsulated magnetic

field sensors based on LaSrMnO thin films, Sensors Actuators A Phys. 228 (2015) 112–117.

[120] Wang Y., Xu Y., Luo L., Ding Y., Liu X., Huang A., A novel sensitive nonenzymatic glucose sensor based on perovskite $LaNi_{0.5}Ti_{0.5}O_3$-modified carbon paste electrode, Sensors Actuators B Chem. 151 (2010) 65–70.

[121] Shahnavaz Z., Lorestani F., Alias Y., Woi PM, Polypyrrole–$ZnFe_2O_4$ magnetic nanocomposite with core–shell structure for glucose sensing, Appl. Surf. Sci. 317 (2014) 622–629.

[122] Nossov A., Rinkevich A., Rigmant M., Vassiliev V., Combined lanthanum manganite magnetoresistive-fluxgate magnetic field sensor, Sensors Actuators A Phys. 94 (2001) 157–160.

[123] Xu Y., Memmert U., Hartmann U., Magnetic field sensors from polycrystalline manganites, Sensors Actuators A 91 (2001) 26–29.

[124] Rubi D., Fontcuberta J., Lacaba M., González AM, Baztán J., Calleja A., On–off magnetoresistive sensor based on screen-printed $La_{2/3}Sr_{1/3}Mn\ O_3$ manganite, Sensors Actuators A Phys. 132 (2006) 52–55.

[125] Balcells L., Calvo E., Fontcuberta J., Room-temperature anisotropic magnetoresistive sensor based on manganese perovskite thick films, J. Magn. Magn. Mater. 242–245 (2002) 1166–1168.

[126] Jung H.-R., Lee S.-G., Kim M.-H., Yeo J.-H., Structural and electrical properties of $Ni_xMn_{3-x}O_4$ system ceramics for infrared sensors, Microelectron. Eng. 146 (2015) 109–112.

[127] Shinagawa M., Kobayashi J., Yagi S., Sakai Y., Sensitive electro-optic sensor using $KTa_{1-x}Nb_xO_3$ crystal, Sensors Actuators A Phys. 192 (2013) 42–48.

[128] Sun LL, Tan OK, Liu WG, Zhu WG, Yao X., Poling of multilayer $Pb(Zr_{0.3}Ti_{0.7})O_3$/$PbTiO_3$ thin film for pyroelectric infrared sensor application, Infrared Phys. Technol. 44 (2003) 177–182.

[129] Shahnavaz Z., Woi PM., Alias Y., A hydrothermally prepared reduced graphene oxide-supported copper ferrite hybrid for glucose sensing, Ceram. Int. 41 (2015) 12710–12716.

[130] Fine GF, Cavanagh LM, Afonja A., Binions R., Metal oxide semi-conductor gas sensors in environmental monitoring, Sensors 10 (2010) 5469–5502.

[131] Filipovic L., Selberherr S., Performance and stress analysis of metal oxide films for CMOS-integrated gas sensors, Sensors 15 (2015) 7206–7227.

[132] Wang Y., Liu B., Xiao S., Li H., Wang L., Cai D., High performance and negative temperature coefficient of low temperature hydrogen gas sensors using palladium decorated tungsten oxide, J. Mater. Chem. A 3 (2014) 1317–1324.

[133] Addabbo T., Bertocci F., Fort A., Mugnaini M., Shahin L., Vignoli V., An artificial olfactory system (AOS) for detection of highly toxic gases in air based on $YCoO_3$, Procedia Eng. 87 (2014) 1095–1098.

[134] Ding J-C, Li H-Y, Cai Z-X, Wang X-X, Guo X., Near room temperature CO sensing by mesoporous $LaCoO_3$ nanowires functionalized with Pd nanodots, Sensors Actuators B Chem. 222 (2016) 517–524.

[135] Wadkar P., Bauskar D., Patil P., High performance H_2 sensor based on $ZnSnO_3$ cubic crystallites synthesized by a hydrothermal method, Talanta 105 (2013) 327–332.

[136] Ghosh P., Mukherjee A., Fu M., Chattopadhyay S., Mitra P., Influence of particle size on H_2 and H_2S sensing characteristics of nanocrystalline nickel ferrite, Phys. E Low-Dimensional Syst. Nanostructures 74 (2015) 570–575.

[137] Itagaki Y., Mori M., Hosoya Y., Aono H., Sadaoka Y., O_3 and NO_2 sensing properties of Sm $Fe_{1-x}Co_xO_3$ perovskite oxides, Sensors Actuators B Chem. 122 (2007) 315–320.

[138] Fan K., Qin H., Wang L., Ju L., Hu J., CO_2 gas sensors based on $La_{1-x}Sr_xFeO_3$ nanocrystalline powders, Sensors Actuators B Chem. 177 (2013) 265–269.

[139] Bhargav KK, Maity A., Ram S., Majumder SB, Low temperature butane sensing using catalytic nano-crystalline lanthanum ferrite sensing element, Sensors Actuators B Chem. 195 (2014) 303–312.

[140] Verma N., Singh S., Srivastava R., Yadav BC, Fabrication of iron titanium oxide thin film and its application as opto-electronic humidity and liquefied petroleum gas sensors, Opt. Laser Technol. 57 (2014) 181–188.

[141] Addabbo T., Bertocci F., Fort A., Gregorkiewitz M., Mugnaini M., Spinicci R., et al., Gas sensing properties and modeling of $YCoO_3$ based perovskite materials, Sensors Actuators B Chem. 221 (2015) 1137–1155.

[142] Ding J-C, Li H-Y, Guo X., CO sensing mechanism of $LaCoO_3$, Solid State Ionics. 272 (2015) 155–159.

[143] Sutka A., Mezinskis G., Lusis A., Jakovlevs D., Influence of iron non-stoichiometry on spinel zinc ferrite gas sensing properties, Sensors Actuators B Chem. 171–172 (2012) 204–209.

[144] Karthick Kannan P., Saraswathi R., Impedimetric detection of alcohol vapours using nanostructured zinc ferrite, Talanta 129 (2014) 545–551.

[145] Zhao L., Steinhart M., Yosef M., Lee SK, Schlecht S., Large-scale template-assisted growth of $LiNbO_3$ one-dimensional nanostructures for nano-sensors, Sensors Actuators B Chem. 109 (2005) 86–90.

[146] Cheng H., Lu ZG, Synthesis and gas-sensing properties of $CaSnO_3$ microcubes, Solid State Sci. 10 (2008) 1042–1048.

[147] Machappa T., Badrunnisa S., Ambika Prasad MVN, Zinc tungstate ($ZnWO_4$) doped polyaniline (PANI) as humidity and gas sensing composite, Sens. Bio-Sensing Res. (2015).

[148] Singh J., Srivastava M., Kalita P., Malhotra B.D., A novel ternary Ni Fe_2O_4/Cu O/Fe O-chitosan nanocomposite as a cholesterol biosensor, Process Biochem. 47 (2012) 2189–2198.

[149] Ensafi AA, Alinajafi HA, Jafari-Asl M., Rezaei B., Ghazaei F., Cobalt ferrite nanoparticles decorated on exfoliated graphene oxide, application for amperometric determination of NADH and H_2O_2, Mater. Sci. Eng. C 60 (2016) 276–284.

Chapter 5

Air-Suspended Silicon Micro-Bridge Structures for Metal Oxide-Based Gas Sensing

Keshavaditya Golla and Eranna Golla

Additional information is available at the end of the chapter

http://dx.doi.org/10.5772/62647

Abstract

Maintaining specific temperature is a key parameter for most of the gas sensing materials, particularly for metal oxide-based thin film layers, to operate them more efficiently to detect different gaseous (or vapor) species at ppm and ppb concentration levels. For field applications, battery-operated micro-gas-sensors, power dissipation and the required temperature stability are to be maintained with a close tolerance for better signal stability and also to analyze the generated data from the sensing element. This chapter mainly focuses on several metal oxide films to highlight the base temperature and its creation on silicon-based platforms. Air-suspended structures are highlighted, and a comparison is drawn between simple and MEMS-based structures from the power dissipation point of view. Ease of fabrication and operation limitations are explained with fabrication issues.

Keywords: Metal oxide thin films, Gas sensing devices, Platinum micro-heater, Platinum resistance thermometer, MEMS, Surface micromachining

1. Introduction

A lot of research effort has been directed toward the development of miniaturized gas sensing devices, particularly for toxic gas (or vapor) detection and also for pollution monitoring in general and industrial ambience. Such devices should allow continuous monitoring of the concentration levels of particular gaseous species in a quantitative and selective way to enable one to take specific remedial action against the prevailing situations. Though various techniques are available for gas detection, studies based on gas chromatography methods are popular for their accuracy in quantifying the species. This is a major chemical analysis instrument for separating species in a complex sample. The system uses a flow-through narrow tube, known as the column, through which different air samples pass in a gas stream at different rates. Depending on their various chemical and physical properties and their interaction with a specific column filling, they get separated and are subsequently detected. For field applications, such instrumentation usage has practical limitations, and hence, portable devices are preferred. In this respect, solid-state metal oxides offer a wide spectrum of materials, and their sensitivities for different gaseous species, thus making it a better choice for miniaturized sensing devices. However, the efforts have not yet reached commercial viability for many gaseous species because of problems associated with the sensor fabrication technologies applied to these micro-gas-sensing devices.

Primarily the gas detection method is based on the changes in the electrical properties of the sensing films, such as electrical resistance (or conductance); in this case, they are thin metal oxide films. The key detection process in these films is the change of the oxygen concentration surrounding the sensing element. This is caused by the adsorption of the test gaseous species and subsequent heterogeneous catalytic reaction of oxidizing and reducing gaseous species to which they are interacting at the surface. The electrical conductivity depends on the gas atmosphere and on the temperature of the sensing material exposed to the test gas. The signal generated from the sensing element strongly depends on the temperature of the element also. Hence, one has to keep the temperature of the element with an external heater such as platinum in most of the cases. Other elements include doped polycrystalline silicon and nichrome (alloy of nickel and chromium) for this application.

Almost all the solid materials interact with the surrounding areas with their exposed surfaces. The physical and chemical compositions of these solid surfaces determine the nature of these interactions. The surface chemistry will influence many properties, and a gas species interaction is one of the key parameter for gas sensing phenomena. In these reactions, only very small proportions of the atoms of most solids, that too very close to surface areas, participate and interact. These interactions also depend upon the shape and surface roughness conditions. With the increasing demand for better gas sensors, of higher interactions and greater selectivity, intense efforts are being made to find suitable solid materials with the required surface and bulk properties useful for these applications. Material properties and the gaseous species structure cum polarity of the gas molecule play a key role at this stage. It is always desired that the material to be used in the fabrication of these sensors do interact with the gaseous species and upon removal of the interacting species the material ability to return to its initial state, which has induced the change, of its electrical properties. Not many sensor materials show this surface property.

2. Metal oxide thin films for sensing applications

The gas sensors using metal oxide semiconductors have been the subjects of extensive investigations for more than four decades primarily focusing on the SnO_2 material. In more recent research efforts, the interest has shifted to some other promising metal oxide thin films, with very interesting detection properties that they exhibit as promising gas sensing materials. Using new metal oxides has several technological advantages, such as: simplicity in element formation and definition of device structure, excellent chemical stability in harsh environment, relatively low material cost for their fabrication, robustness in practical field applications, and adaptability to wide variety of reductive or oxidative industrial ambient conditions. Fairly good studies have been carried out on these groups of gas sensing materials, and at present, it is an active field of research, and many research teams are actively involved for their development. Nano-structured metal oxide films are gaining popularity because of their surface features exhibited by providing additional surface area achievable in these features [1] for more gaseous interactions.

Metal oxide semiconductor gas sensors work mainly on the principle of reversible change of their electrical resistance when exposed to the gas under study. This property is exhibited by most of the sensing materials that are induced by the changes in their chemical environment. It is found that the adsorption of atomic and molecular species on the inorganic semiconductor surfaces affect the surface properties not only related to the electrical conductivity but also influence the material surface potential. Careful study of these variations in the ambient atmosphere and applying them for gas detection scheme will help to analyze the environment to which it is exposed. This change permits detection and nature of the gas and also on the magnitude of change in chemical environment. In almost all the cases, suitability of sensing material, to be used in the fabrication of chemical gas sensor, depends on its ability to return to its initial stage upon removal of cause that induced the change. Metal oxides are better choice in most of the cases to achieve this key point. However, this issue is one of the toughest problems to select the combination of sensing material and the gas/vapor to be detected. Many breakthroughs are expected in this field of research in near future.

As mentioned earlier, the gas (or vapor) detection technique is primarily based on the change in their electrical resistance of the metal oxide films [2] upon their exposure to the test gas under study. The principal detection process is the change of oxygen concentration at the surface of these sensing metal oxides, caused by the adsorption of the test gas and subsequent heterogeneous catalytic reactions of the gas under study. These test gases could be either oxidizing and/or reducing gaseous species. However, these reactions are temperature sensitive and the electrical conductivity (or variation) depends on the nature of the test gas and also on the temperature of the sensing material exposed to this gas. The signal, typically defined as the ratio of resistance variation, generated from the sensing element strongly depends on the operating temperature of the sensing element. It is necessary to keep the temperature of this sensing element, with an external heater, such as platinum, at a predetermined temperature and is generally attached to the sensing element or should be a part along with it. The heating element should be capable of producing higher temperatures than the typical operating temperature ranges of the sensing element during the operation. This excess temperature is necessary and is purpose built and will help for the periodical regeneration of the sensor surface to drive away the unwanted surface adsorbents. The regeneration process is normally carried out at a temperature above the highest temperature at which the sensing devices are typically operated for maximum sensitivity. Typical values are about 50°C in most of the cases. This will drive out other unwanted gaseous species attached to the surface.

3. Importance of sensor temperature

Operating temperature of gas sensors, based on metal oxides, is a key issue and most important parameters for the operation of any sensor. Different oxide materials respond differently to various gaseous species, as shown in **Figure 1** [3], and one has to be very choosy in selecting this value. Most of these sensors operate at above room temperature range extending up to

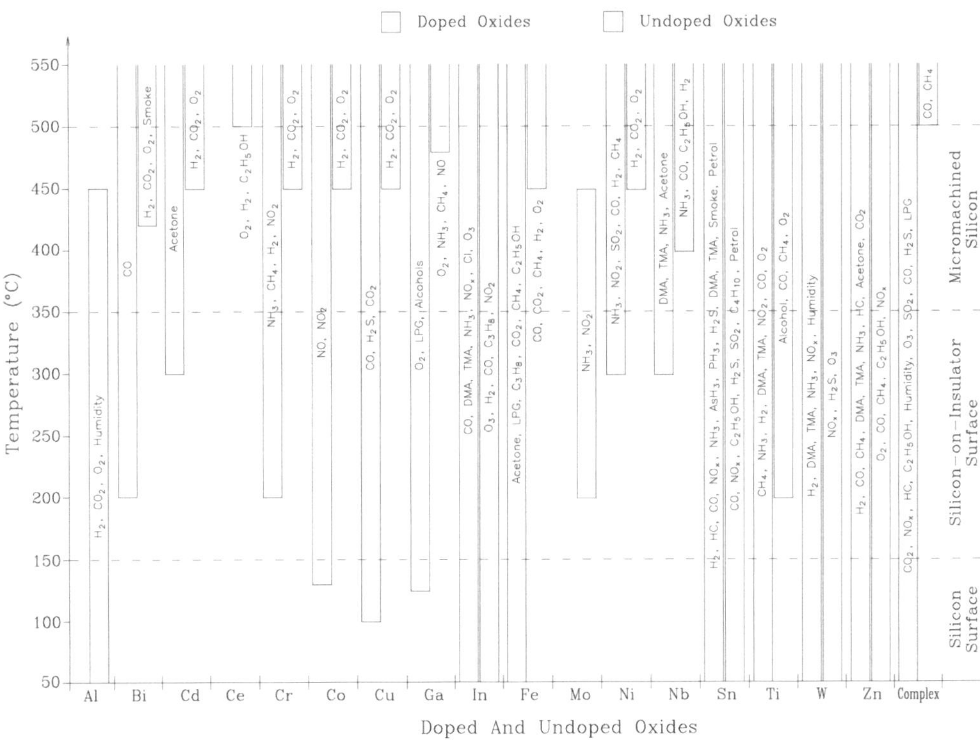

Figure 1. Gas sensing properties of different doped and undoped metal oxides for various gaseous species between the temperatures 50 and 500°C. The regions are placed adjacent to each other. Gaseous species are earmarked to each category. There are three different temperature zones that are identified here to create the required temperatures for the operation of the metal oxide-based sensor devices: on simple oxidized silicon wafer substrate (for operation from 50 to 150°C), SOI wafer surfaces (for operation from 50 to 350°C), and micromachined silicon MEMS structures (for operation from 50 to 500°C). For hybrid microcircuits using alumina substrates, the operating temperature ranges from 50 to 500°C and above ranges [3].

several hundred degrees centigrade, and it is necessary to know the correct operating temperature range of them. These values vary depending on the material and the detecting gaseous species. **Figure 1** shows the gas sensing properties of different doped and undoped metal oxides for various gaseous species between the temperatures 50 and 500°C. To create such temperature, we need a platform where these temperature zones maintain the required operating temperature. Here, three zones are identified for this purpose: simple silicon surfaces (from 50 to 150°C), silicon-on-insulator (SOI) surfaces (from 50 to 350°C), and micromachined silicon structures (up to 500°C). The other option is for hybrid microcircuits, where the operating temperature ranges from 50 to 500°C and above ranges. To obtain good sensitivity, one should know and operate these elements, accordingly.

First, molecular adsorption and desorption are mainly temperature-activated processes and thus decide many dynamic properties of the sensors, such as sensor response time and recovery times, and these parameters depend exponentially on this operating temperature. Surface coverage of the adsorbed species, co-adsorption, chemical decomposition or other reaction issues is also temperature dependent, resulting in different static characteristics of sensors that takes place at different temperature ranges. On the other hand, temperature has an effect on the physical properties of the sensor material, such as charge-carrier concentration, Debye length and work function. In most of the cases, the sensor sensitivity varies with the temperature. Initially, it increases gradually with increase in temperature and becomes less gradual subsequently at higher temperature ranges. With increase in temperature, it exhibits linear dependency initially but later ranges it deviates to non-linearity. This way it attains a maximum sensitivity value, and beyond this point of operation, the sensitivity falls rapidly showing a negative dependency with temperature. This type of behavior is being exhibited by almost all the metal oxide-based gas sensing materials. This is one possible reason for the decreasing sensitivity at higher operating temperatures. Some examples and their response behavior are shown below for most popular metal oxides.

Figure 2 shows the sensitivity graph illustrating the response of SnO_2 sensor to 100 ppm methane gas in dry air as a function of sensor operating temperature [1, 4]. Here, the sensitivity increases continuously from 300 up to 450°C, and this is the best performance point. If one exceeds this operating range, at higher temperatures, there is a negative influence to the sensitivity and it will fall. Subsequently, the sensor sensitivity with temperature changes from positive coefficient to negative coefficient values and degrades the sensor performance. Further, rise in temperature will reduce the sensitivity values considerably as shown in the figure. From the experimental results, one can identify and choose the best operating temperature range for the sensing element by carefully operating the sensor performance with temperature. In this combination, 425–475°C is the best operating zone. It is rare to get such symmetrical behavior of sensitivity values. The slopes and best operating temperature range are identified for this combination. No significant sensing property is possible beyond a specific temperature range, and the values are not good for signal analysis.

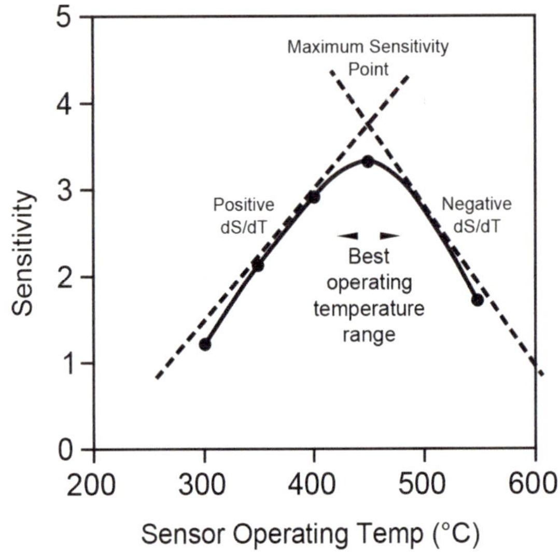

Figure 2. Response of tin oxide-based sensor to 100 ppm of CH_4 in dry air [1, 4] as a function of sensor operating temperature (in °C). Here, the sensor sensitivity, with operating temperature, changes from positive to negative values. Best operating temperature zone for the sensor combination is indicated. No sensing property is possible beyond a particular temperature range, and the recorded values are not good for signal analysis.

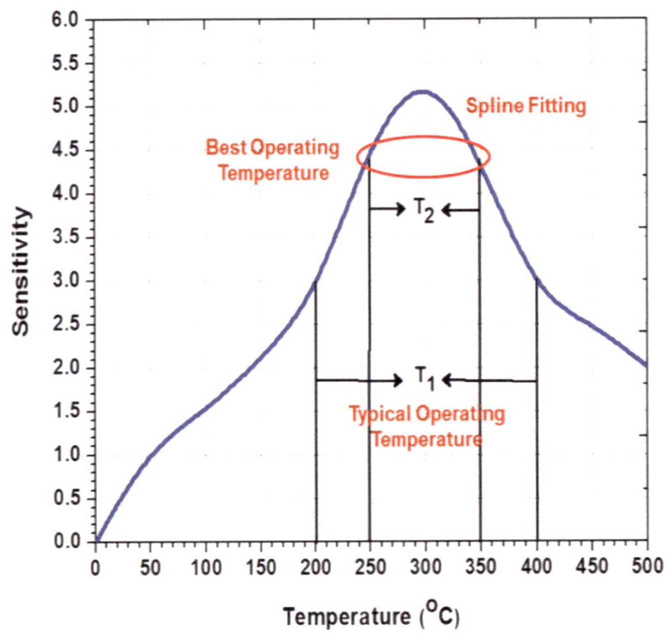

Figure 3. Typical zinc oxide sensitivity vs. temperature range for detecting 200 ppm CO gaseous species [5]. Both operating temperatures and best operating regions are identified for the data recorded using spline fitting. Beyond these ranges, the sensitivity becomes rather poor.

http://dx.doi.org/10.5772/62647

In the **Figure 3**, typical zinc oxide sensitivity with temperature is shown for the detection of 200 ppm carbon monoxide using zinc oxide sensing layer [5]. Both typical operating and best operating temperature regions are identified for the data recorded using spline fitting for the experimental points. Beyond these ranges, the sensitivity becomes rather poor. This material does exhibit sensitivity at lower temperatures, but the curve is not very symmetric in nature. The best operating temperature in this combination is seen to be between 250 and 350°C.

Similar characteristics are seen for the sensors using indium oxide material-based sensor for detecting 2 ppm of nitrogen dioxide gas, as shown in **Figure 4** [1, 4]. This is an important industrial raw material used in the synthesis of nitric acid. In this experimental observation, the sensitivity initially rises with temperature, and after reaching to a maximum value, it falls steadily moving towards a saturation stage after crossing the maximum sensitivity point. Here again, the sensor sensitivity with temperature changes from positive value to negative value. Best operating range is indicated here for sensor operation and is very narrow in these experimental results and is recorded between 165 and 185°C only. Positive slope indicates the reaction limited case, whereas the negative slope is due to binding energy of the gaseous species. At higher temperatures, the gaseous species fail to react with the sensor surface regions and are not detected.

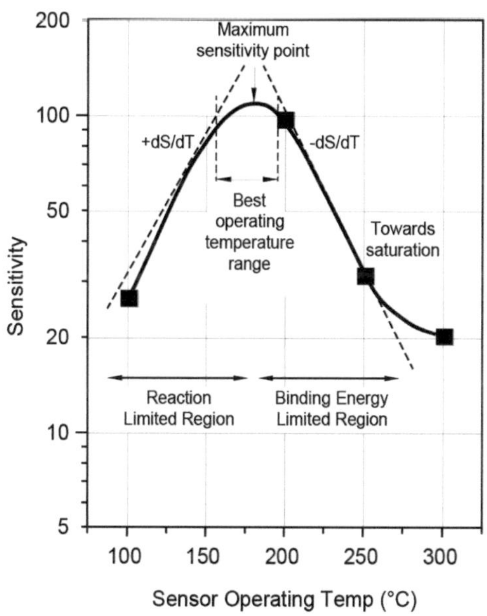

Figure 4. Response of indium oxide sensor to 2 ppm of NO_2 gaseous species as a function of sensor operating temperature (in °C) [1, 4]. Sensor sensitivity changes with the increase in operating temperature from positive value to negative values. Best operating range is indicated here for sensor operation to obtain useful signal for analysis. Positive slope signifies the reaction limited case, whereas the negative slope is due to binding energy of the gaseous species and fail to react with sensor surface regions.

Tungsten oxide also shows similar behavior as indicated in **Figure 5**. Here, the material sensitivity to 1 ppm nitrogen dioxide as a function of operating temperature is shown with spline interpolation of recorded data [6]. Temperature in the range of 135–165°C is the best suited for this combination of gaseous species and sensing material. This is also relatively narrow region. According to Ghimbeu [6], the detection mechanism of nitrogen dioxide has some similarities to that of zinc oxide. It is considered that at low operating temperatures, these gas molecules are adsorbed on the WO_3 surface and forms nitrite type absorbents (ONO^-) and dissociates into nitrosyl-type absorbents (NO^-, NO^+). Hence, the response is related to the catalytic reaction of WO_3 with adsorbed species. The release of electrons from the surface increases the film sensitivity. The maximum sensitivity point here is 150°C.

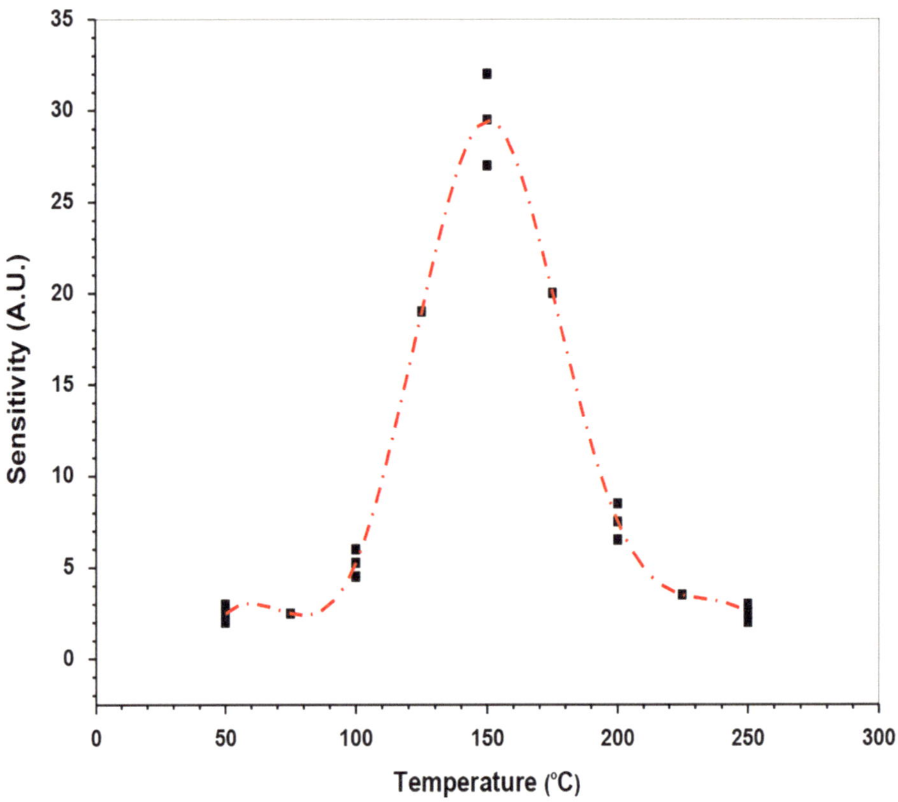

Figure 5. Tungsten oxide sensitivity to 1 ppm NO_2 as a function of operating temperature with spline interpolation of data. Data points are taken from Ref. [6]. Temperature in the range of 135–165°C is the best suited for this combination of gaseous species and sensing material.

4. Microelectromechanical structures (MEMS)

Silicon is the material used to create most of the integrated circuits used in the consumer electronics, and it has a major role in MEMS applications also. For many sensor applications, single-crystal silicon wafer, based on its intrinsic mechanical stability and the feasibility of

http://dx.doi.org/10.5772/62647

integrating sensing and the signal processing electronics on the same substrate, often presents an excellent substrate choice. This is a major point when compared with other substrate materials, such as glass, alumina, and hard polymers. For chemical sensors, the substrate is merely a working platform and its choice is not always straightforward. The silicon material has the highest cost per unit area, but this can be ignored by the small feature size possible in implementing silicon-based platforms. The silicon with its extreme surface flatness and well-established coating procedures often is the preferred substrate, especially for thin film-based sensing elements. The application of silicon device processing techniques to these miniaturized gas sensors results in low heating power, due to inherent low specific heat of silicon as a material, and a smaller spread in the electrical characteristics of silicon devices because of material uniformity in growing the silicon wafers. Owing to the low heat capacity of silicon material, this requires a heating power of the order of few milliwatts per any standard device. Reduction of power consumption is also a key aspect to reduce local thermal stress and also to obtain battery-operated devices for smart sensor applications including the field studies and to use in remote and inaccessible locations. These specific advantages have made silicon rather a convenient substrate for housing different gas sensing materials and to create high temperature zones required for sensing elements.

MEMS structures in silicon are fabricated either by bulk micromachining or by surface micromachining techniques. In the former case, the silicon substrate is etched by either wet chemical solutions or by vapor phase etching. Both anisotropic and isotropic etchants are used for this application. These etching steps are generally deeper and could etch out the silicon wafer depth to a level of 80–90% of the total thickness. The purposes of etchings aim to isolate the part, where high temperatures are to be created. Without these steps, it is difficult to stop the heat flow generated through the heating element mainly because of the good thermal conductivity of silicon and the generated heat spreads very fast into the bulk silicon. The thin platforms thus formed are used for placing the sensing element along with the heating elements. Etching to the predetermined depths in silicon wafer and the screening materials during the silicon for selective etching is very crucial and a challenging task.

In case of surface micromachining, an air gap is created using sacrificial layers. This created air gap prevents the heat flow to the bulk silicon and reduces the body thermal mass. The sacrificial oxides are typically deposited through low temperature CVD process and are etched out at a later stage to create the physical gaps. Many options are available for this purpose to create/deposit the sacrificial layers. Platform structures are selected such that they are thermally stable, mechanically stronger, and HF acid-resistant materials. These are mandatory as the material has to withstand high temperature operation and should not deform during their operation. At the time of etching of the sacrificial oxides, the platform material must show inertness to the etchant. In this category, only two material options are possible. They are LPCVD polycrystalline silicon and LPCVD silicon nitride films. In case of polycrystalline silicon films, it satisfies all the three requirements. HF-based solutions do attack silicon nitride, but the etch rates are very low in comparison with the sacrificial oxides. In this study, both options are used for platform formation and the experimental observations are reported.

Figure 6 shows the SEM image of platform release using sacrificial oxide lateral etching technique for the creation of air gap on silicon surface [7]. For this purpose, a blanket CVD oxide, to a desired thickness, was deposited at low temperature on a fresh silicon wafer. Anchor points are defined selectively using photolithography technique, and the oxide was removed from these portions. Polysilicon was subsequently deposited, and the platform area is defined with photolithography technique again. Unwanted polysilicon portions were etched out and also the CVD oxide underneath it. Platform release experiment was carried out by immersing the complete wafer and was examined periodically for the status sacrificial oxide. In this picture, the platform is not fully released. Since the film is transparent, at this thickness value, the sacrificial oxide is clearly visible through the platform body. The sacrificial oxide regions are totally removed under the supporting arms because of low lateral distances, and the freed supporting structures of platform are visible clearly in the figure. **Figure 7** shows the closer view of the measurements of the created air gap highlighting one of the corner portions of the platform [7]. The vertical measurements were done to confirm the air gaps created in this approach. The slight variation depends on the sacrificial oxide thickness uniformity and also due to errors in marking the regions. Tilt of SEM image capture has its limitations and direct air gap measurements are rather difficult.

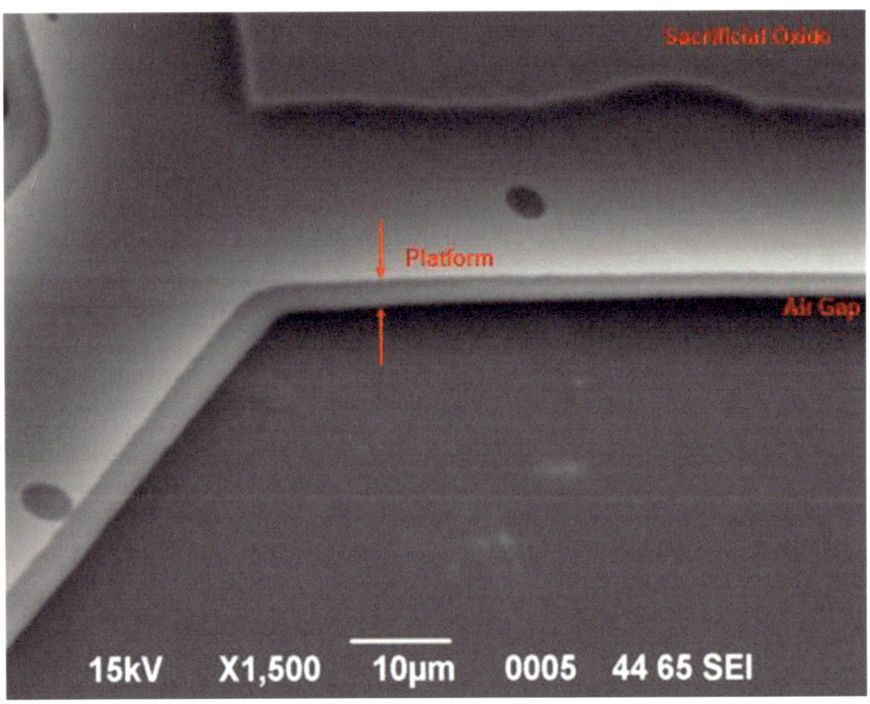

Figure 6. SEM image of platform release using sacrificial oxide lateral etching technique for the creation of air gap on silicon surface [7]. Here, the platform is not fully released. The film is transparent, at this thickness value, and the sacrificial oxide is clearly visible through the platform body, whereas the sacrificial oxide regions are totally removed under the supporting arms because of low lateral distances.

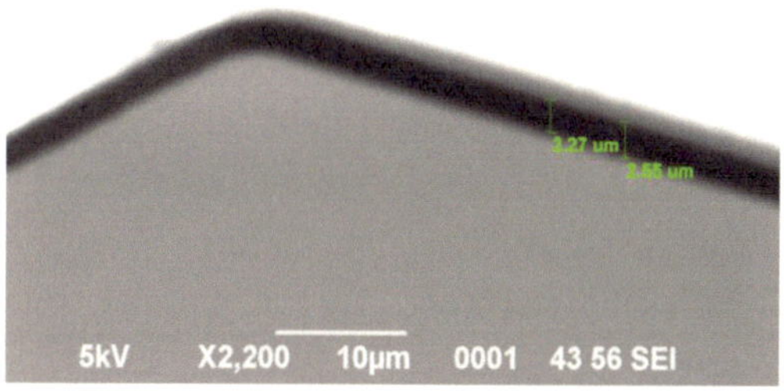

Figure 7. SEM image of platform released by surface micromachining structure on silicon surface highlighting one of the corner portions [7]. The vertical measurements were done to confirm the air gaps created in this approach.

Figure 8 shows the different popular platform structures used for this application. In this test structure, all these shapes are defined and were subsequently etched for releasing the platforms. This figure shows the lateral etching of sacrificial oxide and the platform status of release. The etch front and the released corners are shown in this SEM images. **Figure 9** shows the similar experiment with LPCVD silicon nitride using lesser sacrificial oxide thickness, that is 1.0 μm. These silicon nitride films were densified to reduce the HF acid effect during the platform release. This optical photograph shows the fully released thin film silicon nitride structures.

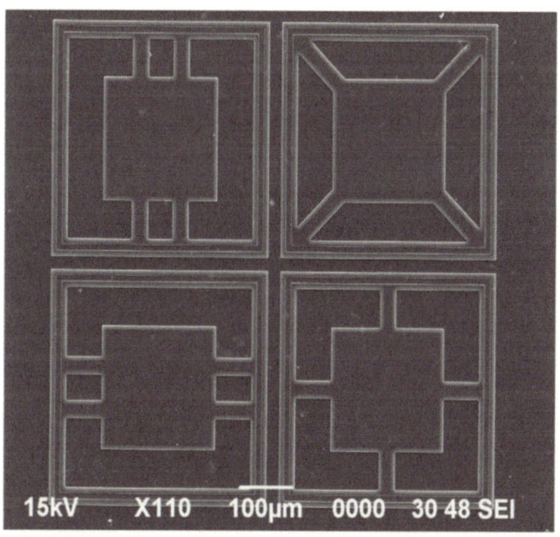

Figure 8. Polycrystalline silicon platform suspension structures for creating an air gap of approximately 3.0 μm [7]. This polysilicon was deposited by LPCVD technique based in silane chemistry. These platforms are in the process of release using surface micromachining technique. Figure shows the lateral etching of sacrificial oxide, and the platforms are semi-transparent at this thickness. The etch front and the released corners are shown in this SEM images.

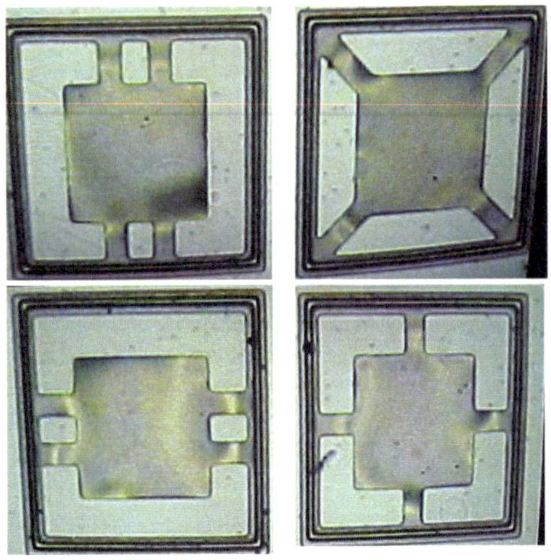

Figure 9. Optical micrographs of different suspended and released silicon nitride platforms by surface micromachining technique on silicon substrates [7]. These nitride films were deposited using LPCVD technique using dichlorosilane- and ammonia-based chemistry. Films were subsequently densified prior to releasing the platform structures. Sacrificial oxide portions are completely removed using hydrofluoric acid-based solutions.

A test run for these release experiments provides enough information to estimate the time durations to complete the entire process. As these structures will reduce the thermal mass considerably and greatly reduces the power dissipation to raise the platform temperatures. They also respond very fast to the changing measurement conditions and exhibit quick thermal response in maintaining the temperatures. The lateral etching and removal of the complete sacrificial layer are time-consuming processes, and the process bound wafer experiences other technology-related issues because of long time contact of hydrofluoric acid. To avoid such time delay, perforated platforms are proposed and many scientific teams are working in this issue. This method reduces the time, but the advantage of perforation limits the mechanical strength of the platform and disturbs the temperature distribution on it. However, there are several practical difficulties in realizing the complete gas sensor devices on these released platforms. Releasing these thin diaphragm platform is a serious technology issue and often decides the yield.

5. Air-suspended structures

To avoid technological limitations, as discussed above, an attempt is made to use the silicon wafer directly as a platform for the required temperature distribution. When compared with the thickness of MEMS platform, say few microns, the bulk silicon wafer thickness is much larger and consumes proportionately larger power. However, the mechanical strength and temperature distributions advantages are not negligible and the device can be put to use for gas sensing applications. As discussed earlier, most of the metal oxide sensors work at higher temperatures, of the order of few hundreds of centigrade, heating the complete silicon die is worth to note. No mechanical issues are reported for silicon in this temperature range and are stable to operate for longer periods. It is also important to note the temperature of the platform to monitor and also to control it to the desired ranges.

In most of the cases, the heating element is made up using platinum metal. This material has been widely used for harsh environment and also known for its standard in many industrial applications. It is chemically inert, and no electrical parameter drifts are reported. In this case, we have used this material and the same is sputter deposited on to the substrate for this application. Standard photolithography and lift-off technique is used for defining the heater for this purpose. Substrates were properly cleaned before the deposition step was carried out. Adhesion layer is used to enhance the adhesion of the film. Defining both the heater and platinum resistance thermometer (PRT) together has certain operational advantages. However, getting a standard PRT is a difficult task and one can define the platinum resistor along with the heater and calibrate it to the desired temperature range is a better option to know the exact temperature value.

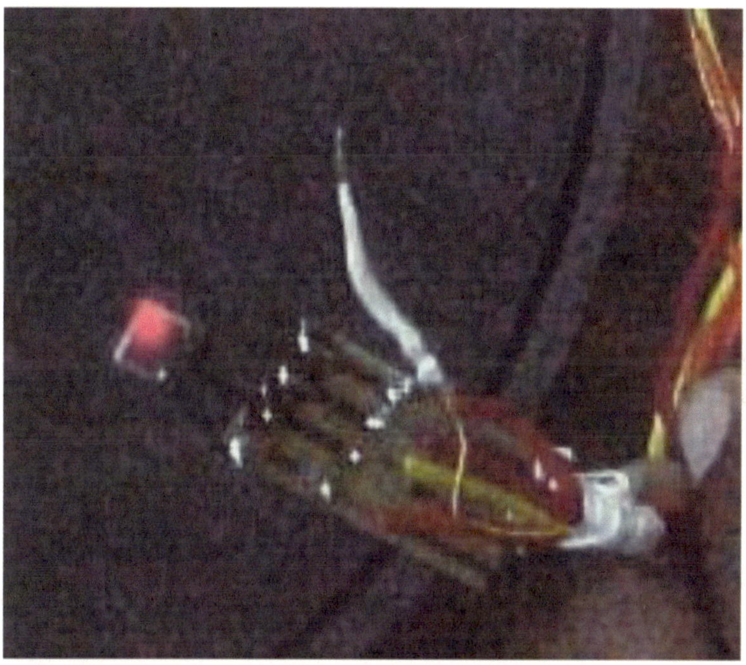

Figure 10. Testing of platinum micro-heater (4 × 5-mm size) on 1-mm-thick alumina substrate up to red hot temperature ranges.

To verify the micro-heater performance, platinum metal was deposited on 1-mm-thick polished alumina substrate and was defined by lift-off technique [8]. Each heater die was measured as 4-mm × 5-mm, and the electrical contacts were made to the heater pads to check the performance. DC current was applied, in steps, to know the heating of the alumina substrate. For temperature measurement purpose, a thermocouple was placed on the back side of the alumina heater to measure the exact temperature directly. Heater current was increased further till the red hot condition was observed on the alumina substrate. **Figure 10** shows the testing of this micro-heater up to red hot temperature ranges. Through this experiment, we gained confidence to operate these heaters to a value not exceeding 500°C, the required temperature for most of the metal oxide-based gas sensing devices.

Figure 11 shows the PRT resistance variation with the measured temperature using a thermocouple attached to the backside of the alumina substrate. The alumina die size is same in this case also. Both the heater and PRT are defined on the same side of the polished alumina substrate, and the electrical contacts were made to them to measure the current rating. Platinum thickness is arbitrarily selected for this study, and we avoided the theoretical prediction of the PRT resistance values. Only DC current, through a constant current supply source, is applied for the heater pads to raise the platform temperature. Variation of PRT resistance and the temperature measured was recorded. The experiment was repeated at different values of heater current (in mA). Both the values are recorded to study the behavior. These experimental observations corroborate the functionality of both heater and defined PRT on the alumina substrate.

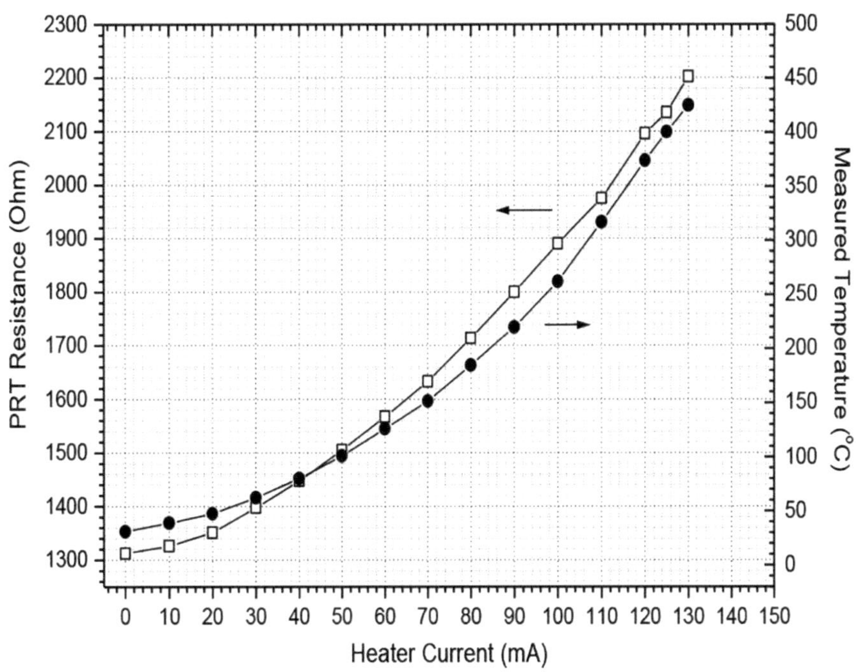

Figure 11. PRT resistance variation with measured temperature using thermocouple attached to the backside of alumina substrate. Data are drawn together to the same experimental points. Measured temperature is recorded through the thermocouple attachment and using standard calibrated system which converts the thermocouple emf values to equivalent temperature. In this case, the alumina substrate is one-mm thick and the die size is 5 × 4-mm defined area. Heater and PRT are defined on the same side of the substrate. Pt thickness is arbitrarily selected for this study, and DC current is applied for the heater pads.

http://dx.doi.org/10.5772/62647

Air-suspended structures suffer from thermal equilibrium issue even though they are defined on the same side on alumina substrate. Heating in these structures draws more power because of larger mass when compared with the thin MEMS platform structures. The thermal equilibrium may take couple of minutes to reach the equilibrium stage. Hence, this thermal response time issue will remain with these structures and that influences the sensor response and recovery time values.

6. Fabrication of sensing element, PRT, and heater structures

The electrical conductivity is a critical issue for the sensing element and systematic measurement of it is very important to the functionality of sensor. The resistance of thin film consists of three parts: (i) bulk resistance, (ii) surface resistance, and (iii) contact resistance between metal electrodes and the oxide-semiconducting films. However, for practical applications, all these three values are considered as a single entity and the total resistance is directly accepted as the sensing element resistance.

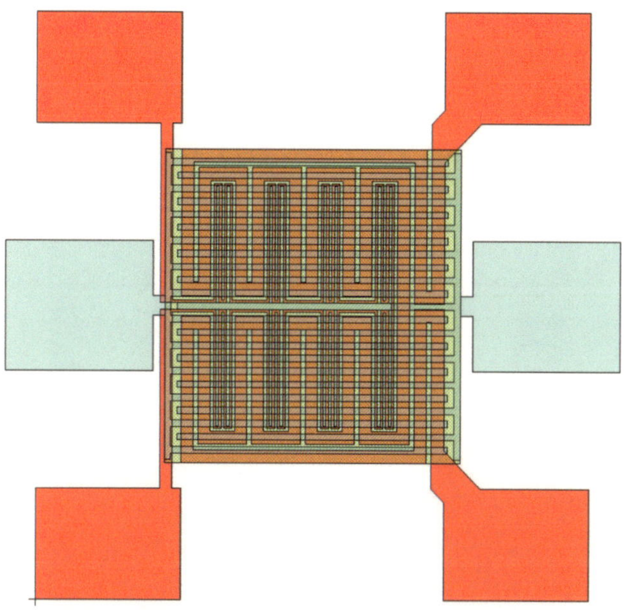

Figure 12. Design for heater and PRT configuration to raise the sensing layer temperature [7]. Both heater and PRT are defined by single photolithography step, and the total area is covered by the sensing film as shown above. Interdigitated structure measures the changes in electrical resistance of the sensing layer. Here, all the layers are electrically isolated with CVD oxides where applicable.

For air-suspended silicon micro-bridge structures, we have designed a structure such that both the PRT and heater components are defined in a single photolithographic step. After the definition of this PRT embedded heater, the total area is covered with a sensing layer to be used for gas sensing applications. This sensing layer will be directly interacting with the atmosphere, where surface reactions take place. To measure the sensing layer conductivity variations, an interdigitated structure is defined and placed directly on this layer. The ideal metal is gold and is preferred because of its inertness in the atmosphere conditions. Electrical isolation is done with low temperature-deposited CVD oxides to bury the heater and PRT structures. Contact windows are opened on the pad areas for wire bonding. **Figure 12** shows the complete design, where PRT, heater, sensing element, and interdigitated structures are designed. For clarity, the CVD isolation oxides are not shown in this figure.

For the deposition of platinum and also for the sensing layer, reactive RF sputtering method is used. The depositions are carried out in an oil-free ultra-high vacuum chamber using suitable target materials as per the thickness requirement. Sputtering is done in the argon plasma environment, and the vacuum conditions are closely monitored. Depending on the requirement of film thickness, a predetermined time and RF power values are estimated and the depositions are carried out accordingly. Sometimes film thickness monitors are used to monitor the growing films. **Figure 13** shows the RF sputtering process inside the high vacuum chamber. **Figure 14** shows the photograph of platinum thin film deposited on oxidized 3-inch-silicon wafer using this technique.

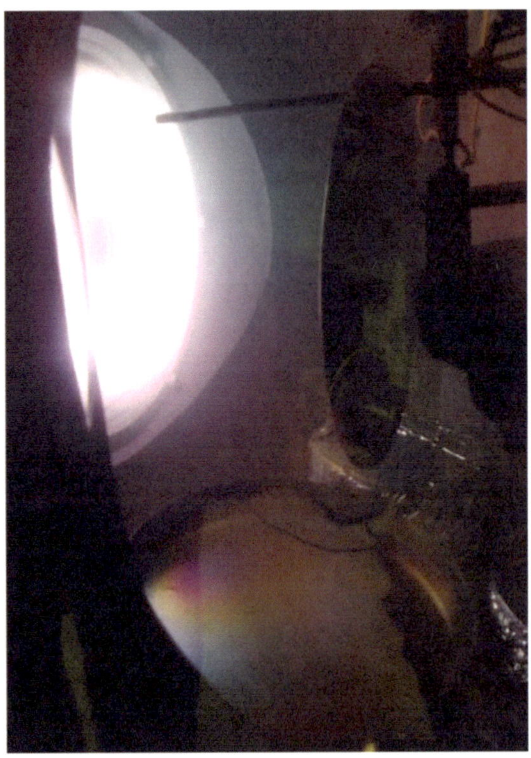

Figure 13. Typical RF sputtering process inside the vacuum chamber, under ultra-high vacuum conditions, to deposit platinum metal used for defining PRT and heater configurations.

http://dx.doi.org/10.5772/62647

Figure 14. Photograph of platinum thin film deposited on oxidized three-inch-silicon wafer by RF sputtering technique.

Figure 15 shows the photograph of PRT embedded micro-heater configuration defined on silicon. This structure is realized using photolithography combined with lift-off technique [8]. The left side of the picture pads shows the PRT connection, whereas the right hand side is meant for the heater configuration. Using this base structure, many sensors can be realized on this. Because of line width difference, the resistance values of PRT and heater vary very widely. Optimization is necessary for the thickness of the platinum film to get calibrated values. To simplify the PRT, a new design was realized to reduce the resistance values very close to the practical values. **Figure 16** shows the simplified version of PRT embedded micro-heater, for a purpose built structure of 5-mm × 4-mm, to create low to moderate temperatures on silicon surfaces particularly suitable for metal oxide-based gas sensing materials [9]. **Figure 17** shows the fabricated devices at wafer level.

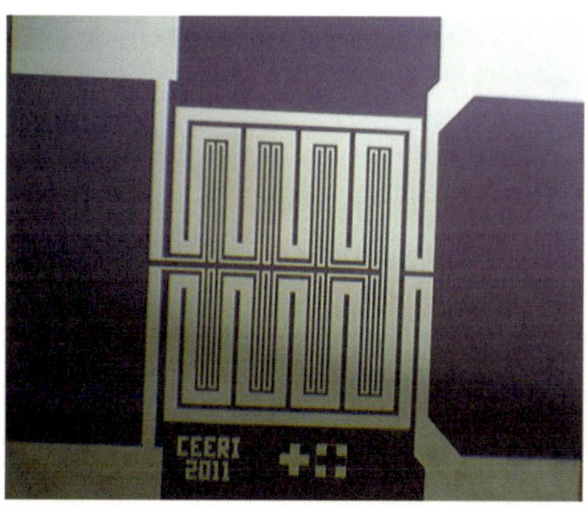

Figure 15. Photograph of PRT embedded micro-heater configuration on silicon. The structure is realized by photolithography and lift-off technique.

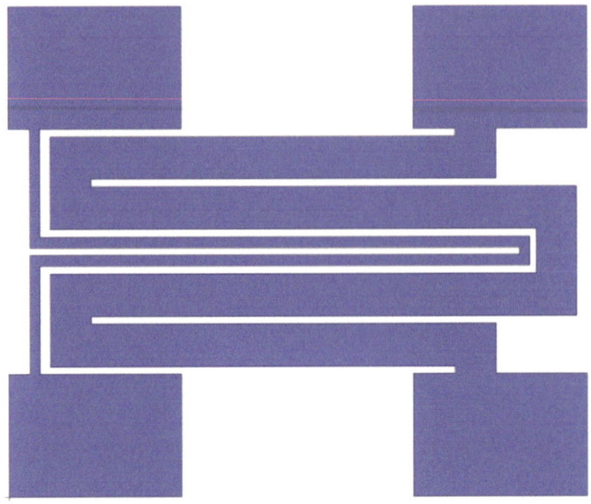

Figure 16. Simplified version of PRT embedded micro-heater, for a purpose built structure of 5 × 4-mm, to create low to moderate temperatures on silicon surfaces suitable for metal oxide-based gas sensing materials [9].

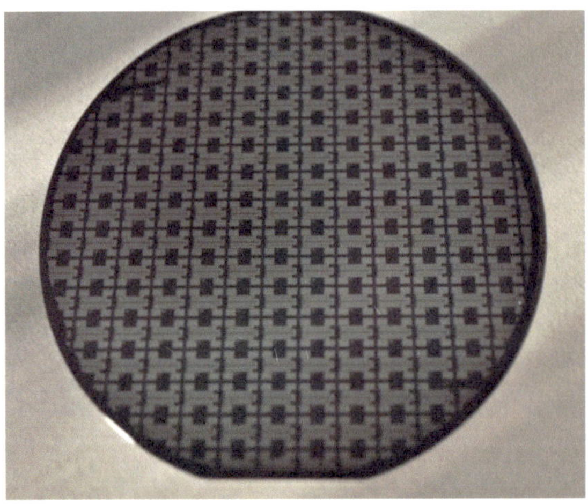

Figure 17. Realization of PRT embedded micro-heater on a 3-inch oxidized silicon wafer [7, 9] using platinum RF sputtering and lift-off methods.

Subsequently, all the dies are separated by standard dicing method and wire bonding was carried out on the defined pad areas. These bonded platforms are connected to a structure to verify the temperature distribution of the micro-heater on these air-suspended micro-bridge structures [9] as shown in **Figure 18**. In this structure, the air acts as a barrier and the heat generated by the heating element initially distributes within the bulk of silicon by thermal conduction process. The heat that escaped through the contact pads is much less when compared to the air convection. This air convection has a greater role in the heat loss mechanism. A closed package housing encloser with proper opening is a possible solution for these proposed structures to avoid convection-related issues.

Figure 18. Wire bonding to verify temperature distribution of the micro-heater on air-suspended silicon micro-bridge structures [9].

Resistance calibration of the platinum PRT is necessary to operate the heater more accurately. It is the PRT value that controls the heater current, and the true values are relevant here. **Figure 19** shows the calibration of PRT (top) and micro-heater (bottom) using standard temperature bath up to 500°C to the structure as per the design shown in **Figure 16**. Here, both the values are stable and are repeatable within the shown range. Figure also shows that the heater will heat up the platform and automatically raises the PRT resistance value signifying the temperature raise. This raises the sensing layer temperature to the required level for the sensor operation.

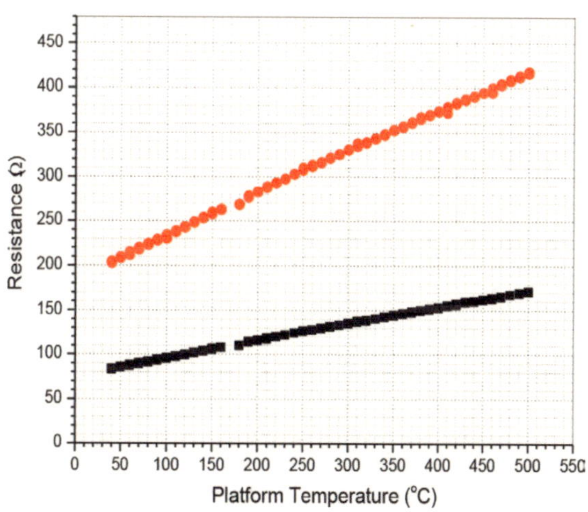

Figure 19. Resistance calibration of the platinum PRT (top) and micro-heater (bottom) using standard temperature bath up to 500°C to the structure as per the design shown in **Figure 16**.

7. Reliability of bulk micro-bridge structures

In most of the cases, the heating element is made up using platinum but polysilicon and nichrome heaters are also reported for this purpose. Not many encouraging publications are found in case of the later materials. Some experiments were carried out with the structure as shown in **Figure 15** in an open environment with minimum air turbulence. **Figure 20** shows the PRT incremental resistance with 1 mA DC constant and continuous heater current with the silicon substrates. Only at the initial stages, there is a rise but within 15 min, the complete die reaches to thermal equilibrium. It is expected that in a closed environment, this time delay may be brought down to couple of minutes. Heating element is placed on one side of the substrate surface and generated heat is slowly distributed into the bulk silicon. This is a time-consuming process.

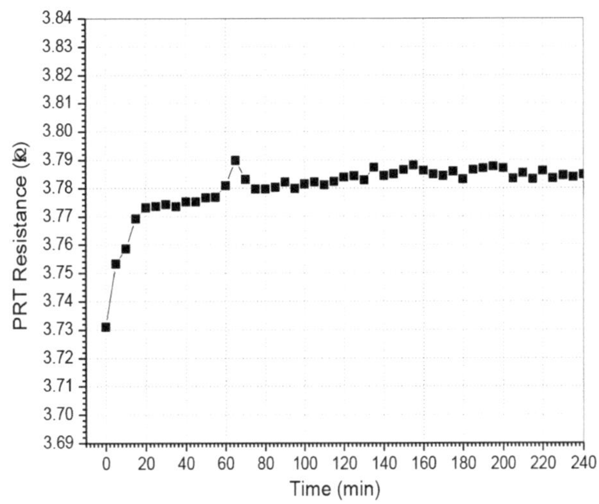

Figure 20. PRT incremental resistance with 1 mA DC constant and continuous heater current with silicon substrates to the structure as shown in **Figure 15** [7].

Thermal hysteresis has been found in these structures because of silicon wafer thickness. **Figure 21** shows the thermal hysteresis measurement of PRT and the heater on oxidized silicon substrate to the design structure shown in **Figure 15**. The circles are during increasing DC current steps, and square points are while decreasing cycle. Here, the current is passed through the micro-heater pad connections. In these recordings, 5 min time gap (delay time) was maintained between apply the heater current and measure the PRT resistance values [7]. The delay is mainly aimed for temperature rise and better uniformity in the bulk of silicon substrate. Despite the delay, a clear thermal hysteresis is seen at 1 mA current setting.

The stability of both PRT and heater was carried out over a period of time, and **Figure 22** shows the results for the structure shown in **Figure 15**. The bottom points correspond to the heater and the top ones to the PRT. These air-suspended silicon micro-bridge structures were carried out in the temperature range between room temperature and 500°C. The structure is very stable for long-time operations up to 400°C, but beyond this range, both PRT and heater values become unstable as shown in the figure. The observed values scatter very randomly. Between 450 and 500°C, the structure becomes mechanically unstable, and ultimately, it fails to provide reliable values. Hence, the entire system is safe to operate up to a value of 400°C. This temperature is sufficient for many sensing materials and cover good number of sensing materials.

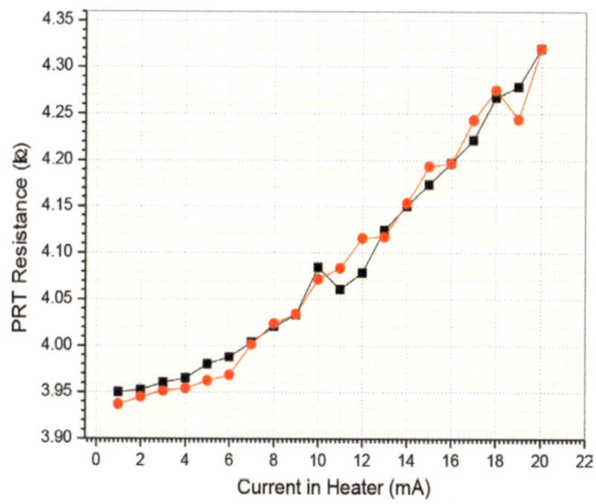

Figure 21. Thermal hysteresis measurement of PRT and heater on oxidized Si substrate to the structure shown in **Figure 15**. The circles are during increasing steps and square while decreasing the current passed through the micro-heater. Here, 5 min time gap was maintained between apply the current and measure PRT resistance values [7]. The delay is mainly aimed for temperature rise and better uniformity in the bulk of Si substrate.

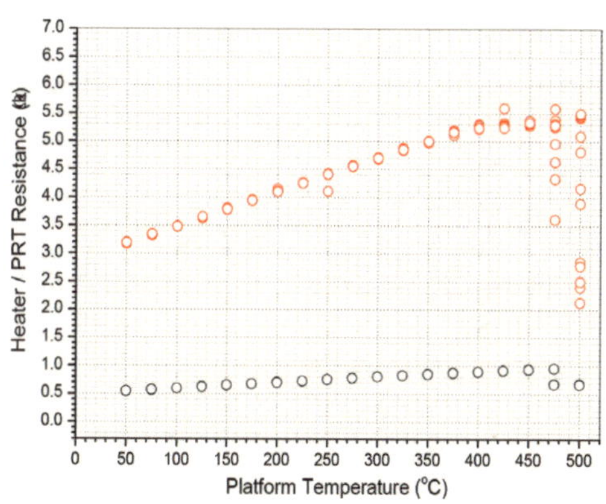

Figure 22. Response of air-suspended silicon micro-bridge structures in the temperature range between room temperature and 500°C. The structure is very stable for long time operations up to 400°C, but beyond this range, both PRT and heater values become unstable as shown. Between 450 and 500°C, the structure becomes mechanically unstable and ultimately it fails.

The structure of heater and PRT configurations can be simulated using ANSYS® software for any specific configurations to achieve the required temperature distribution as per the requirement of sensing film and its physical shape. Necessary physical parameters are to be provided for the realistic simulation. The final temperature distribution, for a specific configuration, as indicated in **Figure 16** is shown in **Figure 23**. Here, the temperature distribution of the micro-heater temperature, operated at 5 V DC applied voltage, is shown. From the results, it is very clear that the center of the suspended platform will have the maximum temperature value. In these locations, the sensing layer needs to be distributed accordingly to enable the element experiences the required temperature for better performance. This way one can select the required temperature and approach the desired properties by applying suitable voltages. This is easy for most of the field applications and to achieve the desired temperature ranges to use the metal oxide-based sensing layers for sensing applications. At this temperature, there are a wide variety of material combinations [9] one can select to detect different gaseous and VOC species. By simulation of temperature and its distribution, it is possible to realize the devices based on air-suspended micro-structures for gas sensing applications.

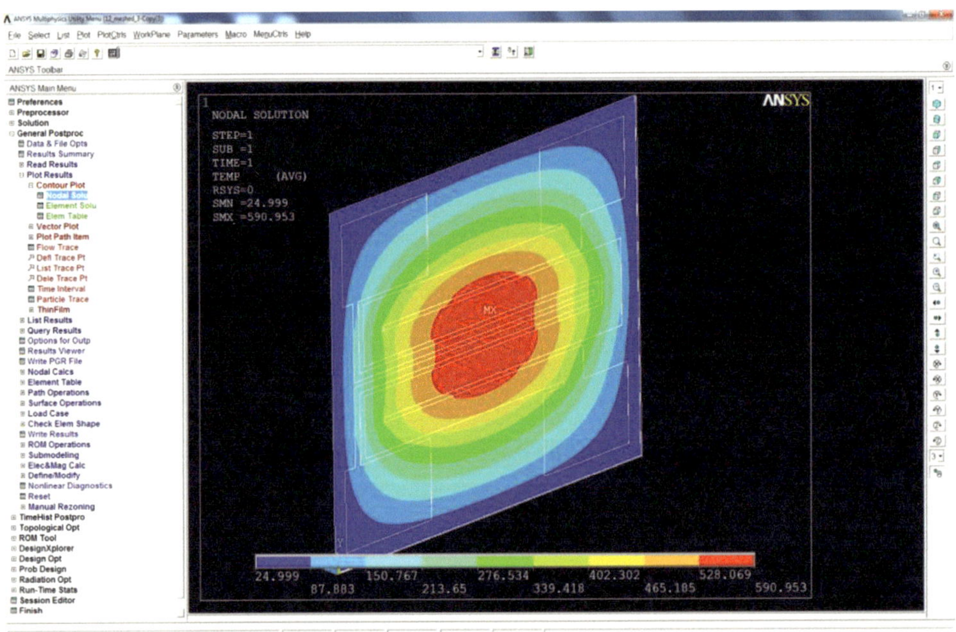

Figure 23. Typical nodal temperature contour plot with 5 V DC input for the structure, as shown in **Figure 16** using battery-operated device. Image is generated by simulation using ANSYS® by providing realistic parameters values [9].

Placing the structures in a sealed environment with a metal mesh opening is a better option to operate these devices. Standard microelectronic packaging is not useful at the heater temperatures. Useful device packages for these micro-bridge gas sensors are proposed in this study. The package structure, as shown in **Figure 24**, which is a standard package used for Toshan LPG gas detector sensor MQ6 [10], appears to be useful. Interaction of atmospheric gases takes place through this metal mesh and the air suspension provides the required thermal isolation from the heating element.

Figure 24. Useful device package structures for the air-suspended micro-bridge gas sensors proposed in this study. Shown here is a package used for Toshan LPG gas detector sensor MQ6 [10]. Courtesy: amazon.in. Source: internet.

8. Conclusions

In this chapter, we mainly focused for creating and maintaining specific temperature ranges useful for most of the gas sensing materials, particularly for metal oxide-based thin film layers, to operate them more efficiently to detect different gaseous species at different concentration levels. For practical devices and also for field applications, battery-operated micro-gas-sensors with air-suspended silicon bulk micro-bridge structures are discussed from the practical usage point of view. This avoids both the bulk and surface micromachined MEMS configurations. Power dissipation and the required temperature stability are discussed to operate the best operating temperature zones for getting a stable and reliable signal. Air-suspended structures are highlighted, and a comparison is drawn between simple and MEMS-based structures from the fabrication point of view. Ease of fabrication and operation limitations are also explained with structure reliability in usage.

Acknowledgements

The authors would like to thank DIT, Govt. of India, New Delhi (DeitY), for project sponsorship at CSIR-CEERI Pilani, India, where some of the experimental work was carried out, and for executing them for VOC gas sensor development for detecting pollutant gases using metal oxide-based thin film structures.

Author details

Keshavaditya Golla[1*] and Eranna Golla[2*]

*Address all correspondence to: keshav.aditya19@gmail.com and eranna.mems@ee.iitm.ac.in

1 Institute for Automation and Electrical Engineering, University of Applied Sciences Bremerhaven, Bremerhaven, Germany

2 Centre for NEMS and Nanophotonics, Department of Electrical Engineering, Indian Institute of Technology, Madras, India

References

[1] Eranna G., Metal Oxide Nanostructures as Gas Sensing Devices, CRC Press (A Taylor & Francis Book), New York, 2012, ISBN 978-1-4398-6340, and the references therein.

[2] Jonda S., Fleischer M. and Meixner H., "Temperature control of semiconductor metal-oxide gas sensors by means of fuzzy logic", *Sensors and Actuators*, B34, pp. 396–400, 1996.

[3] Eranna G., Joshi BC, Runthala DP and Gupta RP, "Oxide materials for development of integrated gas sensors—A comprehensive review", *Critical Reviews in Solid State and Materials Sciences*, 29, pp. 111–188, 2004, and the references therein.

[4] Niederberger M., Garnweitner G., Pinna N. and Neri G., "Non-aqueous routes to crystalline metal oxide nanoparticles: Formation mechanisms and applications", *Progress in Solid State Chemistry*, 33, pp. 59–70, 2005.

[5] Eranna G., Dwivedi M. and Vyas Vimal, "Optimization of ZnO thin film sensor for CO detection using MEMS structures", National Symposium on Materials for Advanced Technology (NSMAT-2011), March 27, Banasthali University, Banasthali, India, 2011.

[6] Ghimbeu CM, "Preparation and characterization of metal oxide semiconductor thin films for the detection of atmospheric pollutant gases", Thesis, Presented at the University Paul Verlaine of Metz, November 2007.

[7] Eranna G., Project on Development of MEMS based integrated micro gas sensor for VOCs and pollutant gases, Grant-in-Aid Project from Department of Information Technology (DeitY), Govt of India, New Delhi.

[8] Eranna G., Joshi BC, Bhargava J., Sharma AK, Kothari P. and Laxmi S., "A new lift-off technique to define fine platinum metal features using CVD sacrificial layers", Patent pending.

[9] Keshavaditya G., Eranna GR and Eranna G., "PRT embedded microheaters for optimum temperature distribution of air-suspended structures for gas sensing applications", *IEEE Sensors Journal*, 15 (7), pp. 4137–4140, 2015.

[10] https://www.google.co.in/webhp?sourceid=chrome-instant&ion=1&espv=2&ie=UTF-8#q=Toshan LPG gas detector sensor MQ6.

Chapter 6

Graphene-Based Junction Devices for Hydrogen Sensors

Sukumar Basu and Surajit Kumar Hazra

Additional information is available at the end of the chapter

http://dx.doi.org/10.5772/62734

Abstract

Graphene is quite a robust material for sensing hydrogen and other gases at room temperature as well as at elevated temperatures with high efficiency. This chapter deals with different junction devices based on graphene for hydrogen sensing. Graphene has excellent electronic attributes that make it suitable for gas sensor devices. However, till date, the research on graphene-based junction devices is not many. In this chapter, we present different types of graphene junction devices suitable for hydrogen sensing. Hydrogen sensor response of these junctions is analyzed, and the sensing mechanism is presented. The temperature- and atmosphere-dependent inversion of n-type to p-type conductivity in graphene is highlighted for hydrogen sensing. Moreover, the two dimensional nature of graphene makes it very convenient for device miniaturization. This chapter provides relevant information on the growth of graphene, the fabrication of different graphene junction devices, and hydrogen sensor applications. Also, the sensor-related concerns such as cross-sensitivity, signal drift, stability, and interference of humidity during hydrogen sensing are thoroughly discussed in this chapter.

Keywords: Graphene, Junction devices, Hydrogen sensor, Wide temperature range, Selectivity

1. Introduction

The importance of hydrogen sensor both domestically and industrially does not need a fresh introduction. Hydrogen has the prime potential applications as fuel in domestic, nuclear installations, and space vehicles. The huge calorific value and non-polluting by-products of hydrogen combustion are two principal advantages of using hydrogen as a fuel. But the negative aspect is its extremely high inflammability at fairly low concentration (~4% and above) when mixed with air leading to disastrous explosions. So, the use of hydrogen gas for combustion or for any other purpose invariably requires a monitoring and controlling device to detect the hydrogen leakage and to alleviate the serious explosion danger. There are continuous R & D efforts to develop hydrogen sensors in different dimensions and for use at different temperatures.

Of late, graphene has been shown to be the most suitable material for hydrogen sensor because of its superior mechanical, electrical, and chemical properties. The two dimensional (2D) structure of graphene is most desirable for the development of miniaturized and fast hydrogen sensor devices. Graphene is a good device material owing to its intrinsic properties. The favorable thermal conductivity (~5000 W/mK) and very high electrical mobility (~200,000 cm^2/Vs) are very special attributes for developing excellent electronic devices [1]. Because of very large surface-to-volume ratio graphene is an interesting gas/vapor sensing material that can allow fast and high gas adsorption. The high hydrogen intake ability of graphene and other carbon nanostructures has been reported in the literature [2–4]. It is also reported with experimental evidences that the high hydrogen intake ability of graphene is mainly due to the intercalation phenomenon [5–7]. The basic concept of hydrogen intercalation in graphene is the attachment of hydrogen with the graphene layer at the graphene–substrate interface. For multilayer structures, the intercalation phenomenon also happens between the graphene layers, and hence, large amount of hydrogen interaction is possible [8–10].

The intercalation occurs between graphene and atomic hydrogen. So, the important concern for hydrogen sensor devices based on graphene is the generation of atomic hydrogen. This is due to the important fact that the percolation of molecular hydrogen through graphene is impossible because of the dimensions of hydrogen molecule and graphene network. The C–C lattice parameter is 2.46 Å, and the size (diameter) of hydrogen molecule is 2.89 Å [7]. Hence, hydrogen molecule cannot possibly percolate through the graphene layer. On the other hand, the Bohr radius of hydrogen atom is only 0.5 Å, which will easily allow the hydrogen atom to pass through the graphene layer, and subsequently getting attached with the graphene layer between the substrate and graphene. This attachment will temporarily isolate the graphene layer from the substrate, resulting in the intercalation phenomenon. The attachment of hydrogen atoms with the carbon atoms is via van der Waals forces, and hence, the hydrogen atoms can easily desorb upon the reduction of hydrogen partial pressure at a particular temperature during the recovery of the sensor signal. The generation of atomic hydrogen in graphene-based sensor devices depends on the activity of the graphene surface. For instance, the presence of functional groups can initiate reaction with hydrogen, or the presence of catalytic metal can trap hydrogen and break the molecules into atomic hydrogen. Sometimes, junctions with another semiconductor can also bring interesting sensing attributes.

In this chapter, we shall highlight the graphene-based devices that have potential for developing hydrogen sensors, with appreciably fast response and recovery characteristics at different temperature regimes. Since graphene is a high-temperature material, it can be used for sensing hydrogen and other gases at room temperature as well as at elevated temperature with high efficiency. Again, graphene is a 2D material and so it is convenient to fabricate miniaturized sensor devices for commercial applications. Another unique property of graphene is the reversible inversion of electrical conductivity from p-type to n-type and vice versa with temperature in both oxidizing and reducing atmospheres. The sensing nature can be inverted with equal efficiency with the variation of temperature and environment. This versatile characteristic is rarely found in other sensing materials leading to the innovation of the new kind of junction devices. Graphene can enable smaller and lighter sensors with endless design possibilities. They are more sensitive, can detect smaller changes in matter, work faster, and eventually less expensive than the traditional sensors. Some graphene-based sensor designs contain a field effect transistor (FET) with a graphene channel. Upon detection of the targeted analyte the current through the transistor changes and sends a signal that can be analyzed to determine several variables.

The potential applications of different forms of graphene (including pristine, B- or N-doped and defective graphene) as gas sensors to detect small gas molecules (CO, NO, NO_2, NH_3, and H_2) have been investigated and confirmed theoretically by using density functional computations. The structural and electronic properties of the graphene–gas molecule adsorption adducts principally depend on the structure of graphene and the molecular gas adsorption configuration. The doped or defective graphene exhibits much stronger attraction than the pristine graphene for the above five gas molecules. While the defective graphene shows the highest adsorption energy for H_2, CO, NO, and NO_2 molecules, the B-doped graphene shows the tightest binding with NH_3. Also, the strong interactions between the adsorbed gas molecules and modified graphene can bring dramatic changes in the electronic properties of graphene. Both theoretical and practical evidences support that the sensitivity of a gas sensor using B-doped graphene is almost two orders of magnitude higher compared to pristine graphene. So, the performance of the graphene-based chemical gas sensors can be largely improved using the appropriately doped or defective graphene [11].

2. Graphene synthesis for device fabrication

Graphene can be synthesized by employing different methods. Exfoliation is an easy method, in which graphene layer is isolated from *highly ordered pyrolitic graphite* (HOPG). The exfoliated graphene layer can be transferred to a substrate for device fabrication. However, the size of the exfoliated layer is not suitable enough to cover the whole substrate wafer for large-scale device production [12].

Single-crystalline silicon carbide wafers can be heated at elevated temperatures to generate the graphene layer [13]. In this case, the silicon atoms escape the wafer leaving the hexagonal carbon network. But the transfer of thermally generated films is a major problem by this method [14].

Chemical vapor deposition (CVD) is a convenient technique to grow graphene films for sensor applications. For CVD growth, the substrates are coated with a metal (say copper), which acts as the catalytic layer [15]. Apart from copper, other metals such as nickel, silver, gold, platinum, and cobalt can be used for graphene growth [16]. Due to the high solubility of carbon in copper, it is often preferred for graphene growth by CVD.

For CVD growth of graphene films, normally thermally oxidized silicon substrate is the obvious choice [17]. Plasma-Enhanced Chemical Vapor Deposition (PECVD) grown SiO_2/Si substrates can also be used. However, the surface quality of thermally oxidized silicon substrates is relatively better in comparison with PECVD substrates. A thin film of copper is deposited on the substrate by e-beam or thermal evaporation technique. The copper layer mainly acts as the catalytic layer which initiates the growth. The precursor is normally a mixture of a carbon containing gas (like CH_4), a reducing gas (like H_2), and an inert/neutral gas (like Ar/N_2) [18]. The important parameters of growth are deposition time, flow ratio of the gases, and the temperature of the substrate. All these parameters need to be optimized to have good device quality graphene films. It is reported that the CH_4 molecules decompose on the catalytic copper surface in the temperature range 900–1000°C [19]. This temperature range can be modulated by suitable choice of the catalytic metal. The flow ratio is regulated very precisely with the help of mass flow controllers, owing to the important fact that if too much carbon containing gases are allowed to flow over the heated substrates, secondary nucleation can initiate haphazard growth, leading to poor film quality, which is not desirable for good device performance. Sometimes, oxygen groups need to be attached to the hexagonal carbon network for some specific applications. In that case, very small amount of oxygen can be incorporated into the flow stream during growth. The number of deposited graphene layers depends on the time of deposition and the activity of the catalytic layer. The thickness of the catalytic layer is another important parameter of growth. This is due to the fact that at elevated temperatures during growth, the fresh copper atoms from underneath diffuses through the bulk copper layer up to the surface for fresh reactions. Hence, optimization of the catalyst layer thickness is also an important priority for graphene growth. Owing to the fact that pristine graphene is insensitive to hydrogen and other gases functionalization of the grown graphene layer is often necessary for sensor application. Hence, oxygen functionalized graphene, for example, graphene oxide (GO) and reduced graphene oxide (rGO) are equally important for developing sensor devices. Functionalization with other groups can impart special device properties. Such functionalized layers are especially useful to develop resistive gas sensors. Apart from oxygen and other organic groups, the catalytic metals (Pd, Pt, etc.) are also useful for metal-decorated graphene gas sensors. The catalytic metal nanoparticles (NPs) can simply be dispersed on graphene surface to improve its gas sensing property. By this process, the response parameters such as sensitivity, selectivity, and time of response/recovery can be modulated. Another technique to improve the gas sensing activity of graphene is to disperse graphene layers (obtained by either exfoliation or CVD) in a polymer base [20, 21]. Such mixed composition can give special material properties suitable for gas sensing and other applications.

3. Sensor device characteristics

The characteristics of the graphene-based sensor device depend on both electronic and catalytic properties of the active layer (s). For instance, if the layer is electronically active but insensitive to detecting gases (like pristine graphene), it is not suitable as sensor material. In that case, the sensitivity is to be developed by functionalization or surface modification. Sometimes, both electronic and sensing properties can be achieved by simple surface modification techniques. In fact, the generated charge carriers during sensing need to flow through the device circuit

http://dx.doi.org/10.5772/62734

in order to show appreciable sensor response. But just after generation, if the carriers are annihilated or trapped by defects the response, even though happening, will not be observed. Surface modification technique takes care of the defect treatment and restores good electronic attributes, in many cases. Such surface modified layers can be used to develop good resistive sensors, where the important factor is how effective is the surface in capturing the analyte gas molecules, and how efficient the material is in transferring the generated charge to the electrode. The time of response gets prolonged if the solid–gas interaction is slow. However, the recovery of the sensor solely depends on how good is the material electronically. Whether the charge carrier annihilation is minimum or negligible the recovery characteristics are likely to be good.

Apart from the catalytic properties of the sensing material the I–V characteristics and junction bias are important for the junction devices. Generally, the forward barrier is low compared to the reverse one. However, for the defective active sensing layers, the reverse characteristic of the junction is far from the nature of the ideal diode. Therefore, the sensor performance may differ for the junction devices with forward and reverse biased cases, and hence needs optimization. The junction ideality factor is also an important parameter for hydrogen sensor applications, and it is necessary to improve the ideality factor to obtain the good device characteristics and superior gas sensor characteristics, respectively.

The operational characteristic of a junction device with respect to the junction barrier needs a close attention. The charge carriers generated due to solid–gas interaction reach the junction interface and modulate the junction capacitance. The magnitude of this capacitance is affected by the variation of the junction bias, which regulates the device barrier both in the forward and in the reverse bias operations. Hence, both forward and reverse bias studies are necessary for finding the optimum device characteristics.

Optical devices can operate at room temperature or below, which is good for hydrogen safety. Reflectance, transmittance, absorption, refractive index, etc. are the parameters that can be monitored in an optical sensor device. Palladium-based optical hydrogen sensors adsorb hydrogen and forms non-stoichiometric PdH_x, which changes the optical properties, mentioned above. Upon hydrogen desorption, the original properties of Pd are restored.

Surface acoustic wave (SAW) devices are also popular for gas sensors and are used for hydrogen detection. A transformation of acoustic signal to electrical and vice versa is done in SAW devices with the help of the piezoelectric materials. So, any change in the properties of these piezoelectric materials will modulate the acoustic signal that travels from the input to the output end. An important parameter that is considered for quantification is mass. So, if a gas sensitive material is coated on a piezoelectric substrate, the mass of this material changes upon gas adsorption and affects the traversing of the acoustic signal from the input to the output end. Hence, the synchronous frequency of the signal, which is inserted at the input end, is liable to change upon gas adsorption.

In all the above devices, the material platform is most important, and the device properties are majorly determined by the properties of the sensing material. For sensor devices, the contamination from moisture and other pollutants can alter the device characteristics with prolonged usage. Such deterioration can be prevented by proper device packaging and temperature treatment of the sensing platform in a cyclic manner (low to high and vice versa) to eliminate volatile contaminants.

4. Hydrogen response of different types of graphene-based sensor devices

The traditional semiconductor devices which are used to fabricate the electronic gadgets can be effectively utilized to design and develop the graphene chemical sensors. Different physical and chemical characteristics of the sensing materials and/or the junctions are modified in presence of trace amounts of gases or chemicals. The sensor response is registered as the change of electrical, optical, and chemical properties. The following graphene-based devices are generally used for sensing hydrogen and other trace amount of chemicals, gases, and vapors.

4.1. Resistive devices

The surface sensitivity of materials is normally harnessed by fabricating a simple resistive-type sensor with interdigitated electrodes or with two parallel electrodes (**Figure 1**). Interdigitated electrodes are preferred when the material conductivity is not uniform throughout the exposed surface.

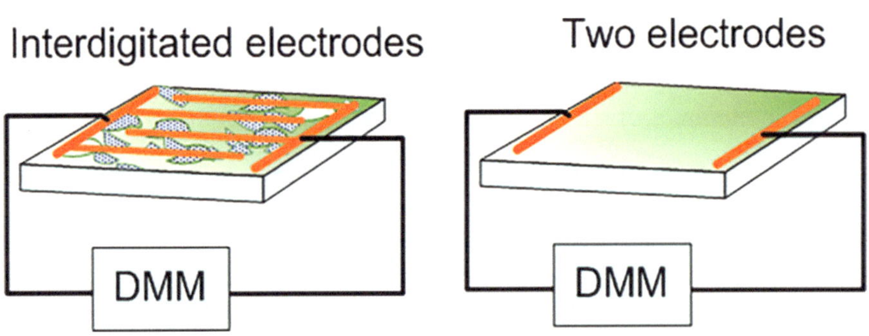

Figure 1. Resistive devices with different electrode configurations.

In case of graphene, pure pristine graphene surface may not be very lucrative to develop simple resistive sensors. However, other forms such as reduced rGO, or surface functionalized graphene, or catalytic NP dispersed on graphene, etc. are useful to develop resistive sensor devices. Zhou et al. [22] reported room temperature operation of resistive sensor devices based on rGO. The flow rates of the test gases influence the adsorption of gases on the active surface, and hence the sensitivity of the resistive device. Also, the rGO thickness is an important parameter for obtaining optimum response characteristics.

Pure platinum (Pt) and bimetallic platinum–iron (Pt_3Fe) NP were used to decorate nitrogen-doped graphene to develop resistive hydrogen sensors [23]. Polyol technique was used to synthesize the NPs and subsequent decoration. The sensors were responsive to 4 vol%

hydrogen and could be used for room temperature sensing. Pure Pt decoration yielded 47% response, while bimetallic Pt_3Fe yielded 35% response in 4% H_2 gas ambient. Pt NP decoration on rGO was also done to develop hydrogen sensors by Ghosh et al. [24]. The response of the devices in H_2/air ambient varied from 19% (200 ppm) to 57% (5000 ppm), and the response time of the devices in air was ~65 s [24]. Similar hydrogen sensor studies with bimetallic-graphene composites are reported [25]. The electrical resistance was measured in 2% H_2/Ar ambient in the temperature range −100 to 100°C. The reported response at room temperature is 4.3%. Optimum response was obtained at 40°C with the response time of 2 s.

Graphene flakes of nanoscale dimension- and spherical-shaped nanoporous palladium particles were combined to make the sensitive platform to fabricate resistive sensors [26]. The sensors were sensitive to hydrogen (10,000–1 ppm) at room temperature. The resistive devices showed repeatable response, recovery and a very high hydrogen selectivity during operation near room temperature. The appreciably good selective response to hydrogen is due to the high affinity of palladium to hydrogen relative to other gases. Also, the relatively lower activation energy for hydrogen adsorption is due to the porous nature of nanosize palladium. As a result, the resistive devices can operate even at room temperature.

Graphene nanoribbons (GNRs) have recently shown the novel physical, electronic, and spin transport properties for the wide range of device applications. Graphene nanoribbon in combination with Pd NPs can also be used for hydrogen sensing [27]. The graphene nanoribbon was fabricated by laser-assisted lithography. The nanoribbon of 200 nm width was patterned at a separation of 1 μm. In an atmosphere of 1000 ppm hydrogen in nitrogen, 90% response was obtained with a response time of 60 s. Almost 80% recovery could be achieved in a time span of 90 s. The devices were also sensitive to other gases. The incorporation of NPs in GNR-based gas sensors can be a novel approach to develop the high-performance gas sensors.

An interesting work on Pd–graphene resistive hydrogen sensor operating with and without bending strain was reported by Yi et al. [28]. A graphene mesh was developed by CVD technique using flexible PET as the substrate. The variation of the resistance of the simple Pd–graphene and the Pd–graphene mesh with bending/unbending cycle was investigated. It was revealed that the simple Pd–graphene showed relatively higher resistance change during bending in air compared to Pd–graphene mesh. The Pd–graphene mesh sensors are more stable to stress for the real field installations with a curvature. At room temperature, the hydrogen response of the Pd–graphene mesh sensor was 1.25 times higher than the simple Pd–graphene in the concentration range 2–15 ppm. Probably, the edges of the graphene mesh provided an additional surface for enhanced hydrogen adsorption and response for hydrogen. The use of flexible substrates was also reported by Chung et al. [29].

4.2. Catalytic dot contact device

The important concern for sensor devices is to attain good response parameters. This can be achieved by reducing the geometrical device area significantly. The geometrical contact area of the catalytic dot contact devices can be made very small. Both planar and vertical configurations (**Figure 2**) are used in this case. The vertical devices are sometimes advantageous because the total vertical transport path of the generated charge carriers is short, which ensures minimum carrier loss due to annihilation and trapping. For transferable graphene films (like those obtained by exfoliation), both planar and vertical devices can be conveniently fabricated. However, for graphene films obtained by CVD on SiO_2/Si substrates (with undoped silicon), the planar device fabrication is more suitable because the insulating SiO_2 is the hindrance for vertical device fabrication. However, if the silicon wafer is highly doped and the barrier SiO_2 layer is very thin, the vertical configurations can be possible to fabricate a catalytic dot contact device for hydrogen sensing.

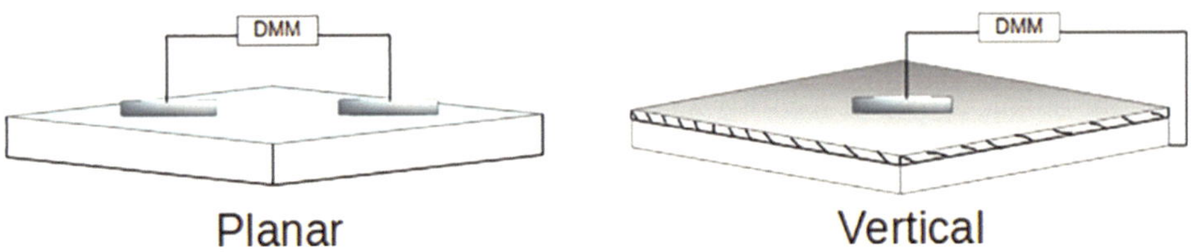

Figure 2. Planar and vertical device configurations.

The reduction in the geometrical area makes the sensor device saturate quickly during response and recover fast due to less adsorption. Hence, the response parameters such as time of response and time of recovery are faster, the essential criteria of an efficient gas sensor. However, the magnitude of % response is relatively less but it is not a big issue in today's sensor research due to advanced electronic circuitry that can amplify signal electronically to get higher electrical output. If the sensing platform is defective in nature, the number of active sites for gas adsorption is higher compared to the defect free matrices. The graphene film grown by CVD is usually defective but it is suitable for gas sensor applications [18, 19]. As reported by Dutta et al. [19], multilayer graphene films on SiO_2/Si substrates showed good response characteristics for hydrogen, with fast response (~12 s). The defective nature of the films was estimated from Raman studies. The nonuniformity of the films was corroborated by Field Emission Scanning Electron Microscope (FESEM) and Atomic-Force Microscopy (AFM) studies. So, considering the defective nature and the uneven surface topography of the CVD grown films, it can be inferred that any catalytic contact on the surface of CVD grown graphene should have large number of active sites for gas adsorption leading to high gas sensitivity. Dutta et al. [19] fabricated Pd/graphene/Pd planar devices using the graphene samples, grown by CVD for 15 min at 1000°C, and obtained good hydrogen sensitivity in the temperature range 110–150°C. The planar device was found to be insensitive to another reducing gas, methane, possibly due to insufficient thermal activation energy to dissociate the large methane molecule. The interesting observation reported by Dutta et al. [19] was the change in the surface morphology after hydrogen exposure. The reorientation of the graphene surface after exposure to hydrogen gas was probably due to the hydrogen passivation of the surface defects. In another study by the same research group using planar Pd/graphene/Pd junctions with graphene grown by CVD for 8 min at 1000°C, hydrogen was detected from room temperature to 150°C [30]. The junctions showed a change in the conductivity from n- to p-type at around 100°C and above, eventually due to hydrogen intercalation that has already been discussed in the introduction part. If current increases upon exposure to hydrogen, it is referred to as n-type sensitivity due to n-type conductivity of the device matrix. And it is p-type response if the current decreases in the presence of hydrogen. At 150°C, the p-type response was reported to be very fast (1–2 s) for the same junction devices, which were very stable in the temperature range 25–150°C. The above discussion supports in favor of using defective graphene thin films as the catalytic contact devices for efficient hydrogen sensing.

Reports are available on the vertical devices with Pt contacts on the surface of graphene grown on SiC substrates [31]. Typical multilayer vertical device geometry is shown in **Figure 3**. Normally, the bottom metal contacts ensure good ohmic metallization, while the top metallic contact is catalytic in nature to accelerate the sensor performance. As reported by Shafiei et al., titanium and platinum multilayer bottom contacts, that is, metal-1 and metal-2 shown in **Figure 3** were used in this study. The graphene films of thickness ~10 nm or less were obtained by the reduction of spray-coated graphitic oxide film. The catalytic metal (Pt) dot (metal-3 in **Figure 3**) accelerated the dissociation of hydrogen molecules during sensing. The devices were operated at 100°C, in an ambient of 1% hydrogen mixed with air. The response was recorded in terms of voltage shift in the dynamic response patterns. At 100°C and in 1% H_2 in air, a 100 mV voltage shift was reported under a current bias of 1 mA.

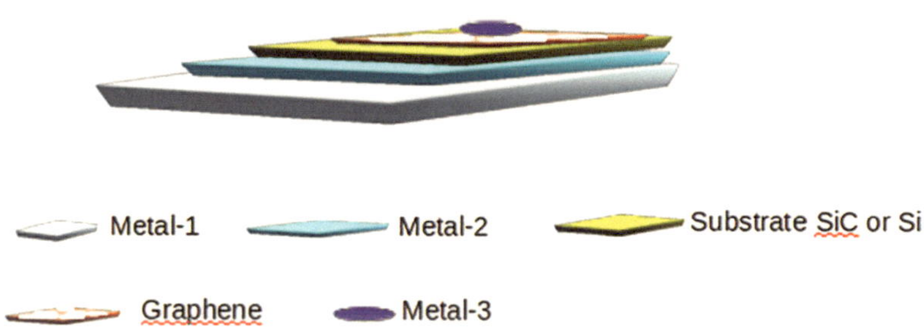

Figure 3. A simple multilayer vertical geometry for graphene-based sensor.

An important condition for both planar and vertical dot contact devices is that the contacts must be ohmic with low contact resistance. Different metal contacts to graphene layers have been studied and reported [32]. The metals selected and reported in this study are Al, Ti, Ni, Cu, Pt, and Pd. The bilayer of either of these metals (10 nm) and gold (50 nm) was deposited on graphene, and the contact resistance was measured. The contact resistance was almost similar in all the cases, and it was improved by different pretreatment. For instance, oxygen plasma treatment of the graphene surface prior to metal deposition gave lower contact resistance when compared with the untreated surfaces or post-processed contacts. The masking is necessary to ensure selected area plasma treatment; otherwise, the graphene surface properties are likely to change. However, Robinson et al. [32] did not report any surface degradation. Post-plasma treatments like thermal annealing at the temperature >450°C, showed a further decrease in contact resistance. A very low value of contact resistance ($4.5 \times 10^{-8}\ \Omega\ cm^2$) was obtained when annealed in the temperature range 450–475°C for 15 min. There is a chance of reduction of the surface carrier concentration upon annealing; however, the carrier mobility is preserved or further improved. Due to removal of volatile impurities from the graphene surface during annealing, the doping level is eventually decreased, thereby reducing the surface concentrations and increasing the purity. So the mobility remains intact or is improved to certain extent. Sometimes, ohmic contacts are used in the vertical devices for hydrogen sensing by displaying the change in the contact resistance. Zhang et al. [33] reported on the transfer of graphene nanoribbon (GNR) onto doped Si substrates with a native oxide layer and decorated with Pd NPs for hydrogen sensor applications. Carbon nanotube (CNTs) unzipped by a simple chemical reduction process was the source of graphene. Also, the synthesis of metal NPs and their incorporation in the unzipped CNTs were done during the chemical reduction. In the vertical structure, silver contact was made on top of the graphene surface and indium at the bottom of silicon substrate. The device sensitivity for 100 ppm H_2 in air was appreciably high to 94% at room temperature. In fact, this vertical structure showed better hydrogen response than the bare and simple resistive Pd–graphene nanoribbon in the identical condition.

A Flow chart for the vertical device fabrication is shown in **Figure 4** below.

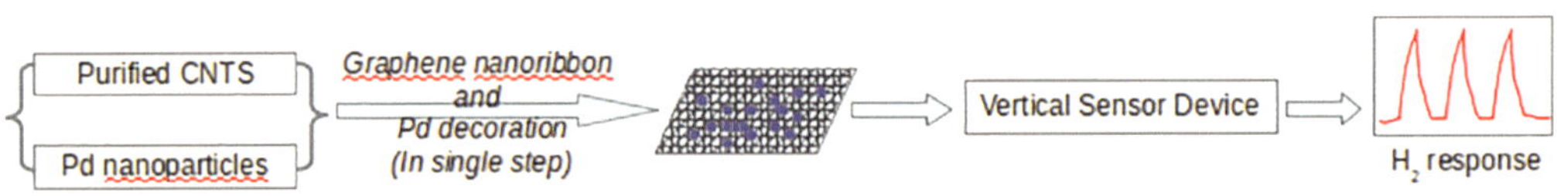

Figure 4. A schematic process flow for the fabrication of Pd-decorated graphene hydrogen sensors.

The improvement in response can be attributed to the concept of enhanced junction sensitivity relative to the simple devices. The vertical junction, (Pd–GNR)/(thin SiO_2 barrier)/(doped p-Si), acts as a simple p–n junction device with the Pd–GNR acting as the n-layer. The junction sensitivity in terms of barrier modulation has been discussed in Section 3 of this chapter. The planar Pd–GNR acts as a simple resistive device, whose surface interactions are similar to the vertical device, but the device speed is low and has relatively higher chances of carrier annihilation due to the planar configuration (also discussed in the beginning of this section). A similar vertical catalytic dot contact Schottky device with rGO (Pd–rGO–SiO_2–Si) was reported by Du et al. [34] with high sensitivity for hydrogen at room temperature.

4.3. Heterojunction device

Junction devices are fabricated with a goal to harness the high-charge carrier mobility of graphene (first layer) in coordination with other attributes of the second layer (either conventional semiconductor like silicon or oxides). Graphene-metal oxide heterostructures are suitable for versatile applications [18, 35–37]. The main advantage of graphene-based heterojunction is the tunable interface barrier, which makes these devices suitable for different applications including chemical sensing as the major area [38]. A basic schematic of a heterojunction device is shown in **Figure 5a**.

Sometimes, both the pristine graphene layer and the conventional semiconductor of the heterojunction are insensitive to a test gas. In that case, the catalytic metals are to be used to

http://dx.doi.org/10.5772/62734

initiate the gas sensor performance. Surface functionalization of the graphene layer can also be an alternative step to improve the sensitivity of the graphene based heterojunction. Normally, graphene-based heterojunction with other gas sensitive materials (like metal oxides, etc.) is likely to show improved sensor performance due to good electronic transport properties of graphene and excellent surface property of the oxide (in terms of gas adsorption/desorption), and scope of interface barrier modulation. The heterojunction can be fabricated by either transferring the graphene layers onto an oxide substrate or CVD grown graphene can be used as a substrate to deposit the oxide film.

A graphene/silicon heterojunction can be biased with pre-optimized bias voltage (either in forward or in reverse bias mode) to perform the sensor studies in a desired gas ambient. Uddin et al. [39] performed hydrogen sensor studies in reverse bias mode with palladium- and platinum-functionalized graphene/silicon heterojunction. The sensitivity of the junction is dependent on the junction bias, which ensures high band bending in the presence of test gas. The graphene/Si heterojunction showed high sensitivity to sub-ppm hydrogen. Liu et al. [40] used CVD grown graphene as substrates to grow ZnO nanowires in order to develop ZnO/graphene nanojunction. These junctions showed hydrogen sensitivity at relatively lower temperatures (<100°C).

A typical hydrogen response behavior of the p-TiO_2/n-graphene heterojunction at different hydrogen gas concentrations and temperatures has been reported by Dutta et al. [18]. The hydrogen response of the p-TiO_2/n-graphene heterojunction in the temperature range 75–150°C, revealed efficient, selective, and stable performance at 100°C. The selective performance was checked by using another reducing gas like methane. Also, an interesting study regarding the device recovery was reported in this work. As shown in **Figure 5b**, the p-TiO_2/n-graphene heterojunction could easily recover back to its baseline value after cutting off H_2 in air and the recovery was incomplete when H_2 was cut off in N_2 ambient. However, the response was similar in both the cases. Hence, it is evident that oxygen plays an important role in the signal recovery process by releasing hydrogen atoms via water formation as shown in Eq. (1) below [18]:

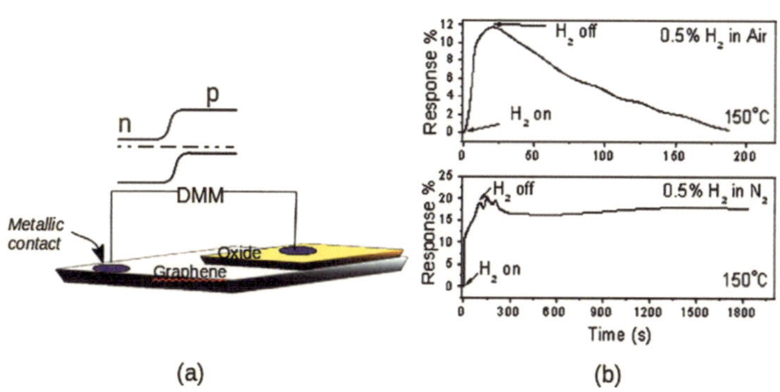

Figure 5. (a) Schematic of a heterojunction device. (b) Comparative sensor recovery in air and nitrogen ambient [18].

$$\left[Pd-(xH^{+})\right]+\frac{x}{4}O_2+xe^{-}\rightarrow\frac{x}{2}H_2O\uparrow+Pd \quad (1)$$

$$\left[Pd-(xH^{+})\right]+xe^{-}\rightarrow\frac{x}{2}H_2\uparrow+Pd \quad (2)$$

Sometimes, the heterojunction configuration also helps in improving the recovery characteristics. While pristine rGO (before and after annealing and with/without UV light exposure) as reported by Kang et al. [41] showed relatively poor recovery characteristics to NO_2, the nano-heterojunction sensor based on rGO and functionalized with tin dioxide nanocluster showed good recovery upon annealing and exposure to UV light. In fact for an optimized tin dioxide functionalization time of 55 min, the nano-heterojunction showed 100% recovery. The good recovery could be attributed to the modulation of the heterobarrier at the nanocontacts between tin dioxide and graphene under UV illumination, which enhanced the rate of surface reactions.

Similar type of nanoscale composite junctions was fabricated with Pd–WO_3 nanostructures and partially rGO by employing a hydrothermal process [42]. This composite material was sensitive to a wide concentration range of hydrogen (20–10,000 ppm) in the temperature range 25–250°C. Simply adding GO to Pd–WO_3 nanostructure showed nearly half the response shown by the nano-heterocomposite between Pd–WO_3 and partially rGO. This could be attributed to the difference in surface area and unsaturated oxygen functional groups between the two, which modulated the surface adsorption. Pd and Pt NPs are sometimes used to decorate heterostructures. As reported by Esfandiar et al. [43], TiO_2/rGO was decorated by Pd and Pt NPs. The reported hydrogen sensitivity of such decorated heterostructures was 92% in 500 ppm hydrogen in air, and the device response time was <20 s at 180°C.

Another type of heterostructures is graphene/Si-nanowire (Si-NW) array for molecular sensing. Such heterostructures are fabricated by using single-layer graphene vertically contacted with high-density Si NWs. Silicon nanowires produced by metal-assisted chemical etching of Si wafer are vertically aligned on CVD grown graphene. Si NWs stand on Si wafer detached from each other under graphene. The uniform Schottky-type junction between Si NWs and graphene thus produced act as the molecular sensors and respond very sensitively to hydrogen gas molecules in air with 1280% resistance change with 0.15 s response/recovery times, the highest performance so far reported. The surface-transfer doping mechanism was suggested based on the results together with the sensor response in vacuum [44].

4.4. Optical device

The response of optical sensors is normally monitored in terms of the parameters such as reflectance, absorbance, refractive index, and photoluminescence [45]. The configuration of a typical optical sensor assembly is shown in **Figure 6**.

http://dx.doi.org/10.5772/62734

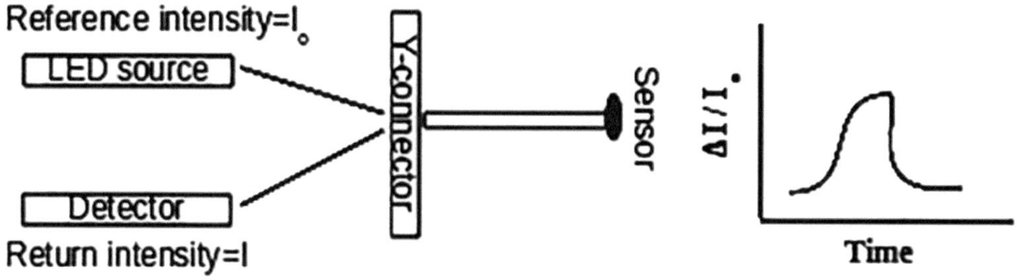

Figure 6. Optical fiber sensor assembly.

The increasing concern of safety demands the room temperature operation of non-electrical gas sensor devices. Hence, the use of optical gas sensors is important. Graphene can be used to develop optical gas sensors for the detection of hydrogen and other gases such as CH_4, NH_3 [46, 47]. For the development of optical gas sensors, graphene has the preference. The chemical stability, affinity to inorganic and organic molecules, scope of surface functionalization and synthesis of composites with catalytic metals and/or other materials make graphene a suitable choice in the field of optical gas sensors [48].

The important step for the development of optical sensors is the transfer of the films from opaque substrates or from substrates with different optical properties. The exfoliated graphene films can be transferred onto any substrate as per the requirement, but for the CVD grown graphene films, the transfer process is complex. Moreover, the substrates must be able to withstand high temperatures because CVD is generally a high-temperature deposition process. Only low-temperature CVD (using good catalytic metals to initiate graphene deposition) is preferred. Lee and coworkers [49] reported on the deposition of graphene films on glass substrates at 450°C using nickel catalytic layer. Although Lee et al. used such low-temperature grown CVD films for humidity sensing, it can also be suitable for hydrogen and other gas sensing. The possibility of surface functionalization can improve the sensing performance. Furthermore, CVD grown films adhere well to the substrates in addition to its stable nature. Quartz substrates can be used instead of glass, if the optical transmission properties of the substrates are to be modulated for gas sensing. So, graphene can also be an appropriate material for the development of optical gas sensor devices.

4.5. Surface acoustic wave (SAW) device

The basic ideas regarding the operation of SAW devices have been discussed in the earlier section of this chapter. SAW is a well-known sensor device, and many sensor studies have been reported. The main concern here is the growth or transfer of graphene films on a piezoelectric substrate. The rest of device fabrication steps, like patterning of electrodes, are simple. SAW resonating platforms based on $LiTaO_3$ have been used for different gas sensing purposes [50]. A typical SAW platform is shown in **Figure 7** below:

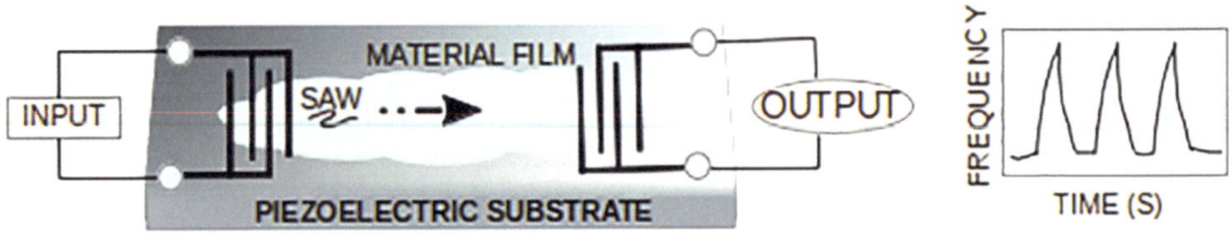

Figure 7. Surface acoustic wave (SAW) device.

The use of graphene-based $LiTaO_3$ SAW device has been reported for hydrogen sensing [51]. The graphene nanosheets were deposited on the $LiTaO_3$ substrate and exposed to hydrogen (0.06–1.0%) in air at room temperature. A typical frequency vs time response pattern was obtained upon hydrogen exposure, from which the response and recovery times were calculated. While the response time decreased with increase in gas concentration, the recovery time increased with increase in gas concentration. Probably, room temperature is not sufficiently enough to ensure complete desorption from the high adsorbing graphene surfaces and therefore, the recovery time is prolonged. Similar devices like graphene-based Rayleigh SAW Resonators fabricated on quartz were used for sensing oxidizing gas like NO_2. The response was good, and the high sensitivity to NO_2 gas was attributed to the defects in the material responsible for the increased gas adsorption [52]. There are also other reports on the use of graphene-based SAW devices for sensing gases like dimethyl-methylphosphonate (DMMP) and dipropylene glycol monomethyl ether (DPGME) are available [53].

4.6. Prospective graphene-based devices

Two grapheme-based prospective devices for hydrogen sensing.

4.6.1. Graphene p–n homojunction

Since 2D graphene can be doped n- or p-type like other semiconductors the graphene p/n homojunction can in principle be designed and fabricated using the traditional semiconductor processing technology. Although quite a few reports on the doping of graphene are available in the literature [54, 55], reports on the use of graphene p/n homojunction as hydrogen sensor is not yet available. The main technical problem during the formation of such junction is eventually the extremely thin dimension of the graphene geometry. Therefore, the technology must be very advanced and sophisticated to realize a successful and sustainable graphene p–n junction. Once succeeded, the graphene p–n homojunction can develop the most efficient hydrogen sensors with miniaturized dimension.

4.6.2. Metal-decorated 2D graphene/MoS_2 flexible wearable gas sensors

The investigation on the gas-sensing properties of a metal-decorated device comprising two-dimensional (2D) flakes of molybdenum disulfide (MoS_2) and graphene electrodes has been reported. This hybrid structure was used to sense ~1 ppm NO_2 gas and ~10 ppm NH_3. Such metal-decorated sensors operate with the mechanism of chemical as well as electronic sensitization. The metal NPs could harness the electronic properties of the 2D graphene/MoS_2 device and enhance the sensitivity for a specific gas molecule.

This work cited an example of NH_3 sensing by graphene/MoS_2 using palladium NPs via accumulation of hole carriers and NO_2 detection using aluminum NPs through depletion of hole carriers. The synergistic combination of metal NPs and the 2D hybrid layers of both graphene and MoS_2 could be utilized for application of this sensor structure as a flexible gas sensor. It was also demonstrated that this hybrid sensor structure did not show any sign of degradation in sensing before/after 5000 times bending cycles. So, this highly sensitive and endurable gas sensor could be used as the wearable electronic sensing device. It was also demonstrated that the metal-decorated 2D graphene/MoS_2 hybrid structure could be effectively used as a selective gas sensor. So, the hybridization of metal NPs and 2D materials could develop a highly sensitive, selective, reliable, and flexible gas sensor by appropriate selection of the metal NP for decoration [56]. The same device configuration could be strategically employed to develop a selective hydrogen sensor.

5. Device concerns of graphene-based gas sensors

5.1. Signal drift and device calibration

An important issue with gas sensor devices is the signal drift with time, and it is a matter of concern for the reliable operation of the sensors. The baseline parameter (either resistance or current) can vary with time, and this is not good for device compatibility with the control electronic circuit. There is also high chance of generating a false alarm. Such drifting can solely be attributed to the material interaction with the atmosphere. Normally, sensing surfaces are very active, and hence can have constant adsorptive/desorptive interaction with the operating atmosphere, which is normally oxidative air ambient (with some humidity concentration). And if the sensing material is defective, which is sometimes an added advantage for gas sensors, such interactions are pronounced. This can lead to change in the baseline resistance with time. Of course, a saturation of the drift is likely with time. However, the process is very slow, and hence, the change is sometimes unpredictable. Moreover, the ambient temperature, if it varies, can change the adsorption/desorption eventually. As a result, the device needs proper calibration both to counter the drift and to have operation in the linear regime of the device characteristics. The calibration is reported for hydrogen sensors [57]. In reality for any device, the linear operational regime is found by plotting the measured parameter (either resistance or current or voltage) with gas concentration. If the plot is throughout linear, the device will have excellent operation at all gas concentrations. However, if the plot is nonlinear, the device needs to be operated in the concentration range where it has linear characteristics.

5.2. Humidity

Humidity is an important concern, and the interference of humidity during sensing other gases is very crucial for all the chemical sensor devices including the graphene sensors. The interference of moisture in the operation of graphene-based sensor devices can be realized and explained in the following manner. The water molecules coming in contact with the graphene surface can interact in two different ways at two different regions of the sensor surface. First, the exposed pure graphene surface (devoid of any catalytic material) attracts polar water molecules and makes hydrogen bonds (one or more hydrogen bonds). During sensing such graphene-adsorbed water complexes transfer very little charge to the active region of the surface, that is, the junction. As an example, the magnitude of charge transfer can be realized from the reported data, that for single walled CNTs (with tube-water separation of 2.69 Å), the charge transfer is of the order of 0.035e, where "e" is the electronic charge [58]. So, the device sensitivity might not be influenced much due to this interaction. However, if the charge exchange is of higher magnitude, the device sensitivity may increase and this is possible if the surface is modified with the functional groups. And for most of the resistive sensors, pristine graphene needs to be modified in some way or other.

The second type of interaction with moisture can happen on the portion of the surface where catalytic materials (like palladium, etc.) are present. On the catalytic surface, water can ionize into H^+ and OH^- and H^+ is likely to be preferentially adsorbed due to the high affinity of catalytic materials (like palladium) to hydrogen. Eventually, the adsorbed H^+ species can increase the concentration of hydrogen at the interface. On the other hand, if the operating temperature is not very high, the OH^- species can occupy/block some of the active sites of the catalytic surface, and the device sensitivity is likely to be affected. Also the sites blocked by OH^- species can prevent further adsorption of the incoming H_2 molecules followed by its dissociation and release of electrons. As a result, the device sensitivity can be affected. So, the presence of moisture can initiate either or both of the above two mechanisms and influence the response characteristics.

An important aspect is the operating temperature during sensor measurements. If the temperature is high (≥100°C), the moisture molecules will easily desorb, while at low temperatures, the interference is likely to be predominant. Therefore, considering all the factors discussed above, the relevant studies are necessary by varying the relative humidity from low (~30%) to high (~90%) to evaluate the moisture interference at the optimum temperature of sensing, for developing a reliable sensor in the commercial scale.

5.3. Stability and selectivity

The reliability and the robustness of the graphene-based gas sensors can be established from the stability studies. The devices are kept under a constant bias for a very long period either in the ambient of a mixture of test gas in air or in simply aerial atmosphere, and the magnitude of the signal (either resistance or current or voltage) is monitored with time. This study can be made continuously for a number of days, and the cumulative data are analyzed. Normally, the stable signal value recorded in the presence of the test gas for a period of one week or so can ensure the stability criteria of an activated sensor. Of course, the background signal without the test gas needs to be checked and calibrated from time to time. The stability study can also be conducted in a discontinuous manner by switching off the sensor device after each day operation and reactivating (by setting in the proper temperature and bias) the next day. Such studies are necessary for the commercial deployment of a reliable sensor device. Dutta and coworkers [18, 19] have reported such studies for graphene-based hydrogen sensors (**Figure 8**).

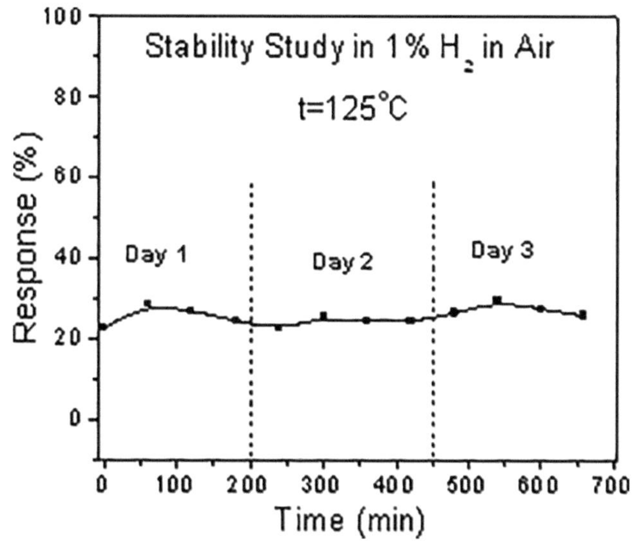

Figure 8. A typical stability study with graphene-based sensor [18].

The selectivity is a pertinent sensor parameter and a matter of concern for gas sensor devices. Almost all the renowned laboratories of the world working on the gas sensor devices consider the selectivity problem a major hurdle against the development of a complete gas sensor. Various methodologies are tried to sort out the problem of cross-sensitivity. Like the well-established oxide-based gas sensors graphene-based gas sensors also face cross-sensitivity problem.

Normally, the problem of cross-sensitivity is tackled by the following sensor operation factors:

1. Operating temperature
2. Gas concentration
3. Material characteristics and catalytic activity.

The operating temperature is a convenient way to eliminate the cross-sensitivity of other gases. For instance, in the case of sensing hydrogen in a mixture with another reducing gas like methane, the lighter gas like hydrogen can be detected at relatively lower temperature compared with methane, a heavier gas, because the heavier molecules require relatively higher activation energy for dissociation and sensing. Hence, some devices can be selective to hydrogen at relatively lower temperature. In the studies performed by Dutta et al. [18, 19, 30] with Pd/graphene and graphene-based heterojunction (**Figure 9**) for hydrogen sensing, it was shown that the devices were sensitive to H_2 but insensitive to CH_4 in the temperature range 75–150°C. This establishes the fact that hydrogen sensing operating temperature does not provide the required activation energy to dissociate heavy molecules like CH_4.

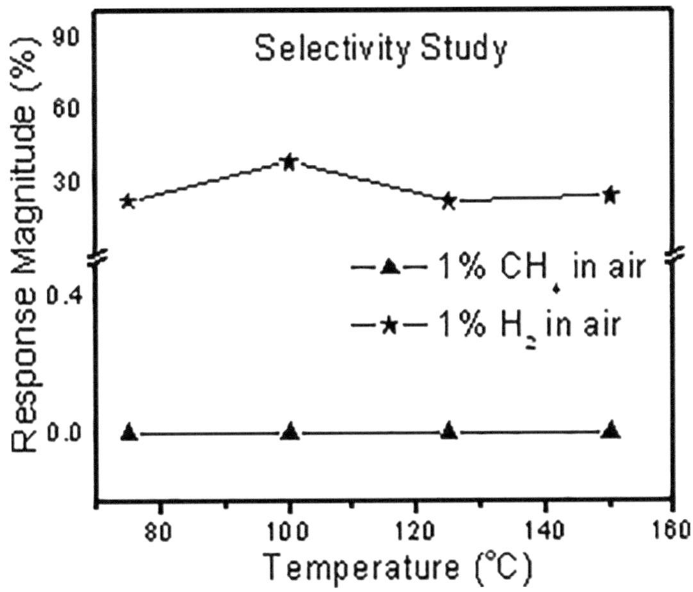

Figure 9. The selective response of TiO_2/graphene heterojunction sensor toward hydrogen gas [18].

The concentration of the sensing gas can determine the device selectivity. Basically, the sensitivity (or % response) is proportional to the gas concentration or the partial pressure of the test gas during sensing. If the partial pressure of the gas is high, the response magnitude becomes high. In the presence of two or more gases, it is likely that the gas component with higher concentration will have relatively higher response at the optimum temperature. In other words, if the gas concentration of the interfering gas (es) is lowered the response magnitude becomes negligible. Then, the device will have dominant sensitivity to the particular gas of interest showing the maximum response. The third important factor is the nature of the sensor surface and its catalytic potential. For example, pristine graphene surface may not be effective for sensing all kinds of gases. So, the surface activity can be improved by functionalization using the catalytic metals, such as palladium, platinum, or by physical/chemical treatment. For instance, studies with pure graphene and Pd–graphene nanocomposite were carried out for sensing hydrogen (1%), moisture, and NO_2 (38 ppm) in a mixture [59]. The response magnitude to 38 ppm NO_2 was excellent with pure graphene but poor with Pd–graphene nanocomposite. On the other hand, the response to 1% H_2 was excellent with Pd–graphene and poor with pure graphene. Moreover, the response was relatively faster for Pd–graphene device due to the catalytic influence of Pd. Furthermore, the presence of moisture influenced the response of Pd–graphene more in comparison with pure graphene. The obvious conclusion from the above study is that palladium-containing graphene sensors will be very good for hydrogen detection; pristine graphene (or functionalized with oxygen species) will be very useful for detecting oxidizing gases, like NO_2. In case, other reducing gases are to be detected, suitable choice of catalytic metal and operating temperature are necessary. The influence of moisture has to be carefully estimated by separate studies in order to have more assuring results.

6. Conclusions and future outlook

Graphene has very high potential for chemical gas sensor applications due to its excellent transport properties, superior mechanical properties and immense scope for the modulation of its surface property by functionalization [60]. Since the surface catalytic property influences the solid–gas interaction, surface functionalization is a convenient technique to improve the response characteristics of graphene-based sensors. The use of catalytic metals such as palladium, platinum, nickel can emphasize the manifold response characteristics and the selectivity. The graphene synthesis methods also play a vital role to increase the device compatibility, and the growth of good quality graphene films requires careful studies with the growth parameters in order to be useful for practical hydrogen and other gas sensors. The various types of graphene-based sensor devices and their sensing mechanisms, particularly for hydrogen sensing, have been discussed in this chapter. The discussions are directed toward understanding the potentiality of the graphene-based devices and their huge promise as a gas sensing platform. The gas detection limit of graphene hydrogen sensors may vary from low to high, which is again a device-dependent characteristic. Normally, catalytic surfaces can detect very low concentration of gases at relatively low temperatures. The importance of defects is quite important for chemical gas sensing, and it is evident from the studies reported so far that the defective surface has high capacity for adsorption, the first step of gas sensing. However, recovery of the sensor signal can be perturbed to some extent due to the gas trapping in the defective sites and it needs relatively higher temperature to remove the adsorbed gases. The multilayer graphene (MLG) hydrogen sensors have normally high response and fast response time at room temperature/moderately high temperature, but the recovery is delayed. However, high-temperature sensing registers faster recovery.

Issues like drift, calibration, stability, selectivity, and the interference of humidity of the graphene-based sensor devices have been thoroughly discussed. These concerns are elaborately considered with special emphasis on the techniques to avert the problems with a simple solution. Sometimes, very simple and judicious choice of some parameters (like temperature) can solve complicated cases of device selectivity in mixed gas ambient. Till date, cross-sensitivity has been an evergreen and challenging problem of the gas sensors in general and the graphene-based devices for hydrogen sensors in particular.

Due to progress in graphene-based research, it is likely that new synthesis techniques may allow the growth of complex device structures, suitable for developing ultra fast sensor devices with selective response characteristics. The problems of suitable ohmic metallization can be addressed if low-temperature CVD techniques are successfully devised. And this will simplify the process steps to fabricate vertical and planar category devices. Also, the quality of the graphene layer can be improved via innovative growth methodologies. These will eventually solve recovery related issues of the graphene-based gas sensor devices. Moreover, it is of utmost importance to understand the operational mechanism of the graphene device at the basic level in order to improvise the ways to improve the performance of the graphene-based sensor devices.

Author details

Sukumar Basu[1*] and Surajit Kumar Hazra[2]

*Address all correspondence to: sukumarbasu@gmail.com.

1 IC Design and Fabrication Centre, Department of Electronics and Telecommunication Engineering, Jadavpur University, Kolkata, India

2 Department of Physics and Materials Science, Jaypee University of Information Technology, Waknaghat, Himachal Pradesh, India

References

[1] Basu S., Bhattacharyya P. Recent developments on graphene and graphene oxide based solid state gas sensors. Sens. Actuators B. 2012; 173:1–21.

[2] Wang Y., Wu Y., Huang Y., Zhang F., Yang Y. Ma Y.F., Chen Y.S. Preventing graphene sheets from restacking for high-capacitance performance. J. Phys. Chem. C. 2011; 115(46):23192–23197.

[3] Fan Z.J., Yan J., Zhi L., Zhang Q., Wei T., Feng J., Zhang M.L., Qian W.Z., Wei F. A three-dimensional carbon nanotube/graphene sandwich and its application as electrode in supercapacitors. Adv. Mater. 2010; 22:3723–3728.

[4] Aboutalebi S.H., Yamini S.A., Nevirkovets I., Konstantinov K., Liu H.K. Enhanced hydrogen storage in graphene oxide-MWCNTs composite at room temperature. Adv. Energy Mater. 2012; 2(12):1439–1446. doi: 10.1002/aenm.201200154.

[5] Riedl C., Coletti C., Starke U. Structural and electronic properties of epitaxial graphene on SiC(0001): a review of growth, characterization, transfer doping and hydrogen intercalation. J. Phys. D Appl. Phys. 2010; 43:374009.

[6] Watcharinyanon S., Virojanadara C., Osiecki J.R., Zakharov A.A., Yakimova R., Uhrberg R.I.G., Johansson L.I. Hydrogen intercalation of graphene grown on 6H-SiC (0001). Surf. Sci. 2011; 605(17–18):1662.

[7] Waqar Z. Hydrogen accumulation in graphite and etching of graphite on hydrogen desorption. J. Mater. Sci. 2007; 42:1169–1176.

[8] Chambers A., Park C., Baker R.T.K., Rodriguez N.M. Hydrogen storage in graphite nanofibers. J. Phys. Chem. B. 1998; 102(22):4253–4256.

[9] Stan G., Cole M.W. Hydrogen adsorption in nanotubes. J. Low Temp. Phys. 1998; 110:539–544.

[10] Arellano J.S., Molina L.M., Rubio A., Alonso J.A. Density functional study of adsorption of molecular hydrogen on graphene layers. J. Chem. Phys. 2000; 112:8114–8119.

[11] Zhang Y.-H., Chen Y.-B., Zhou K.-G., Liu C.-H., Zeng J., Zhang H.-L., Peng Y. Improving gas sensing properties of graphene by introducing dopants and defects: a first-principles study. Nanotechnology. 2009; 20(18):185504.

[12] Wei D., Liu Y. Controllable synthesis of graphene and its applications. Adv. Mater. 2010; 22:3225–3241.

[13] Emtsev K.V., Bostwick A., Horn K., Jobst J., Kellogg G.L., Ley L., McChesney J.L., Ohta T., Reshanov S.A., Röhrl J., Rotenberg E., Schmid A.K., Waldmann D., Weber H.B., Seyller T. Towards wafer-size graphene layers by atmospheric pressure graphitization of silicon carbide. Nat. Mater. 2009; 8:203–207.

[14] Reina A., Jia X., Ho J., Nezich D., Son H., Bulovic V., Dresselhaus M.S., Kong J. Large area, few-layer graphene films on arbitrary substrates by chemical vapor deposition. Nano Lett. 2009; 9:30–35.

[15] Sun J., Lindvall N., Cole T.M., Angel K.T.T., Teng W., Teo K.B.K., Chua D.H.C., Johan L., Yurgens A. Low partial pressure chemical vapor deposition of graphene on copper. IEEE T. Nanotechnol. 2012; 11:255–260.

[16] Batzill M. The surface science of graphene: metal interfaces, CVD synthesis, nanoribbons, chemical modifications, and defects. Surf. Sci. Rep. 2012; 67(3–4):83–115.

[17] Dutta D., Hazra A., Das J., Hazra S.K., Lakshmi V.N., Sinha S.K., Gianonchelli A., Sarkar C.K., Basu S. Growth of multilayer graphene by chemical vapor deposition (CVD) and characterizations. J. Nanomater. Mol. Nanotechnol. 2014; 3(5):004; doi: 10.4172/2324-8777.S1-004.

[18] Dutta D., Hazra S.K., Das J., Sarkar C.K., Basu S. Studies on p-TiO_2/n-graphene heterojunction for hydrogen detection. Sens. Actuators B. 2015; 212:84–92.

[19] Dutta D., Hazra A., Hazra S.K., Das J., Bhattacharyya S., Sarkar C.K., Basu S. Performance of a CVD grown graphene based planar device for hydrogen gas sensor. Meas. Sci. Technol. 2015; 26:115104.

[20] Salavagione H.J., Martínez G., Ellis G. Recent advances in the covalent modification of graphene with polymers. Macromol. Rapid Commun. 2011; 32:1771–1789.

[21] Zheng W., Shen B., Zhai W. Surface functionalization of graphene with polymers for enhanced properties. InTech Publications. 2013. doi:10.5772/50490, Croatia.

[22] Zhou Y., Jiang Y., Xie T., Tai H., Xie G. A novel sensing mechanism for resistive gas sensors based on layered reduced graphene oxide thin films at room temperature. Sens. Actuators B. 2014; 203:135–142

[23] Sripada R., Parambath V.B., Baro M., Nair S.P.N., Sundara R. Platinum and platinum–iron alloy nanoparticles dispersed nitrogen-doped graphene as high performance room temperature hydrogen sensor. Int. J. Hydrogen Energy. 2015; 40(32):10346–10353.

[24] Ghosh R., Santra S., Ray S.K., Guha P.K. Pt-functionalized reduced graphene oxide for excellent hydrogen sensing at room temperature. Appl. Phys. Lett. 2015; 107(15): 153102.

[25] Kumar R., Varadani D., Mehta B.R., Singh V.N., Wen Z., Feng X., Müllen K. Fast response and recovery of hydrogen sensing in Pd–Pt nanoparticle-graphene composite layers. Nanotechnology. 2011; 22(27):275719-1–275719-7.

[26] Phan D.-T., Chung G.-S. A novel nanoporous Pd–graphene hybrid synthesized by a facile and rapid process for hydrogen detection. Sens. Actuators B. 2015; 210:661–668.

[27] Pak Y., Kim S.-M., Jeong H., Kang C.G., Park J.S., Song H., Lee R., Myoung N., Lee B.H., Seo S., Kim J.T., Jung G.-Y. Palladium-decorated hydrogen-gas sensors using periodically aligned graphene nanoribbons. ACS Appl. Mater. Interfaces. 2014; 6(15):13293–13298.

[28] Yi J., Kim S.H., Lee W.W., Kwon S.S., Nam S.W., Park W.I. Graphene meshes decorated with palladium nanoparticles for hydrogen detection. J. Phys. D. Appl. Phys. 2015; 48(47):475103.

[29] Chung M.G., Kim D.-H., Seo D.K., Kim T., Im H.U., Lee H.M., Yoo J.-B., Hong S.-H., Kang T.J., Kim Y.H. Flexible hydrogen sensors using graphene with palladium nanoparticle decoration. Sens. Actuators B. 2012; 169:387–392.

[30] Dutta D., Hazra S.K., Das J., Sarkar C.K., Basu S. Temperature and hydrogen gas dependent reversible inversion of n-/p-type conductivity in CVD grown multilayer graphene film. J. Electron. Mater. (2016); 45. doi:10.1007/s11664-016-4381-0.

[31] Shafiei M., Spizzirri P.G., Arsat R., Yu J., Plessis J.D., Dubin S., Kaner R.B., Kalantar-zadeh K., Wlodarski W. Platinum/graphene nanosheet/SiC contacts and their application for hydrogen gas sensing. J. Phys. Chem. C. 2010; 114(32):13796–13801.

[32] Robinson J.A., LaBella M., Zhu M., Hollander M., Kasarda R., Hughes Z., Trumbull K., Cavalero R., Snyder D. Contacting graphene. Appl. Phys. Lett. 2011; 98:053103.

[33] Zhang Z., Xue Q., Du Y., Ling C., Xing W. Highly enhanced sensitivity of hydrogen sensors using novel palladium-decorated graphene nanoribbon film/SiO_2/Si structures. J. Mater. Chem. A. 2014; 2:15931–15937.

[34] Du Y., Xue Q., Zhang Z., Xia F. Great enhancement in H_2 response using graphene-based Schottky junction. Mater. Lett. 2014; 135:151–153.

[35] Shin K.S., Jo H., Shin H.J., Choi W.M., Choi J.Y., Kim S.W. High quality graphene-semiconducting oxide heterostructures for inverted organic photovoltaics. J. Mater. Chem. 2012; 22:13032–13038.

[36] Riazimehr S., Bablich A., Schneider D., Kataria S., Passi V., Yim C., Duesberg G.S., Lemme M.C. Spectral sensitivity of graphene/silicon heterojunction photodetectors. Solid State Electron. 2016; 115:207–212.

[37] Xia K., Zhan H., Gu Y. Two-dimensional graphene heterojunctions: the tunable mechanical properties. Carbon. 2015; 95:1061–1068.

[38] Bartolomeo A.D. Graphene Schottky diodes: an experimental review of the rectifying graphene/semiconductor heterojunction. Phys. Rep. 2016; 606:1–58.

[39] Uddin M.A., Singh A.K., Sudarshan T.S., Koley G. Functionalized graphene/silicon chemidiode H_2 sensor with tunable sensitivity. Nanotechnology. 2014; 25:125501. doi: 10.1088/0957-4484/25/12/125501.

[40] Liu J.W., Wu J., Ahmad M.Z., Wlodarski W. Hybrid aligned zinc oxide nanowires array on CVD graphene for hydrogen sensing. In: Solid-State Sensors, Actuators and Microsystems (TRANSDUCERS & EUROSENSORS XXVII), Transducers & Eurosensors XXVII: The 17th International Conference, June 16–20, 2013. Barcelona, Spain. p. 194–197. doi:10.1109/Transducers.2013.6626735.

[41] Kang I.-S., So H.-M., Bang G.-S., Kwak J.-H., Lee J.-O., Ahn C.W. Recovery improvement of graphene-based gas sensors functionalized with nanoscale heterojunctions. Appl. Phys. Lett. 2012; 101:123504–4.

[42] Esfandiar A., Irajizad A., Akhavan O., Ghasemi S., Gholami M.R. Pd-WO_3/reduced graphene oxide hierarchical nanostructures as efficient hydrogen gas sensors. Int. J. Hydrogen Energy. 2014; 39(15):8169–8179.

[43] Esfandiar A., Ghasemi S., Irajizad A., Akhavan O., Gholami M.R. The decoration of TiO_2/reduced graphene oxide by Pd and Pt nanoparticles for hydrogen gas sensing. Int. J. Hydrogen Energy. 2012; 37(20):15423–15432.

[44] Kim J., Oh S.D., Kim J.H., Shin D.H., Kim S., Choi S.-H. Graphene/Si-nanowire heterostructure molecular sensors. Sci. Rep. 2014; 4:5384.

[45] Korotcenkov G., Cho B.K., Narayanaswamy R., Sevilla F. Optical and fiber optic chemical sensors. In: Korotcenkov G., editors. Chemical Sensors: Comprehensive Sensor Technologies, Vol. 5, Electrochemical and Optical Sensors. New York: Momentum Press; 2011. pp. 1–56.

[46] Borriello A., Sansone L., Luca G.D., Manna P.L., Musto P., Giordano M. Graphene oxide-based nanohybrid for label-free optical sensing. AIP Conf. Proc. 2014; 1599:489. doi:10.1063/1.4876885.

[47] Gautam M., Jayatissa A.H. Gas sensing properties of graphene synthesized by chemical vapor deposition. Mater. Sci. Eng. C. 2011; 31(7):1405–1411.

[48] Zhu A.Y., Cubukcu E. Graphene nanophotonic sensors. 2D Mater. 2015; 2:032005; doi: 10.1088/2053-1583/2/3/032005.

[49] Lee C.S., Cojocaru C.S., Moujahid W., Lebental B., Chaigneau M., Châtelet M., Normand F.L., Maurice J.-L. Synthesis of conducting transparent few-layer graphene directly on glass at 450°C. Nanotechnology. 2012; 23:265603. doi: 10.1088/0957-4484/23/26/265603.

[50] Chivukula V.S., Daumantas Č., Kim J.H., Rimeika R., Xu J.M., Shur M.S. Surface acoustic wave response to optical absorption by graphene composite film. IEEE Trans. Ultrason. Ferroelectr. Freq. Control. 2012; 59(2):265–270.

[51] Arsat R., Breedon M., Shafiei M., Kalantar-zadeh K., Wlodarski W. (2008). Graphene-like nano-sheets/36° $LiTaO_3$ surface acoustic wave hydrogen gas sensor. In: IEEE Sensors Conference. 2008. Lecce. pp. 188–191. doi:10.1109/ICSENS.2008.4716414.

[52] Thomas S., Cole M., Luca A.D., Torrisi F., Ferrari A.C., Udrea F., Gardner J.W. Graphene-coated Rayleigh SAW resonators for NO_2 detection. Procedia Eng. 2014; 87:999–1002.

[53] Sayago I., Matatagui D., Fernández M.J., Fontecha J.L., Jurewicz I., Garriga R., Muñoz E. Graphene oxide as sensitive layer in Love-wave surface acoustic wave sensors for the detection of chemical warfare agent simulants. Talanta. 2016; 148:393–400.

[54] Meng X., Tongay S., Kang J., Chen Z., Wu F., Li S.-S., Xia J.-B., Lia J., Wu J. Stable p- and n-type doping of few-layer graphene/graphite. Carbon. 2013; 57:507–514.

[55] Joucken F., Tison Y., Fèvre P.L., Tejeda A., Taleb-Ibrahimi A, Conrad E., Repain V., Chacon C., Bellec A., Girard Y., Rousset S., Ghijsen J., Sporken R., Amara H., Ducastelle F., Lagoute J. Charge transfer and electronic doping in nitrogen-doped graphene. Sci. Rep. 2015; 5:14564; doi:10.1038/srep14564.

[56] Cho B., Yoon J., Lim S.K., Kim A.R., Choi S.-Y., Kim D.-H., Lee K.H., Lee B.H., Ko H.C., Hahm M.G. Metal decoration effects on the gas-sensing properties of 2D hybrid-structures on flexible substrates. Sensors. 2015; 15(10):24903–24913; doi:10.3390/s151024903.

[57] Hubert T., Boon-Brett L., Black G., Banach U. Hydrogen sensors—a review. Sens. Actuators B. 2011; 157:329–352.

[58] Zhao J., Buldum A., Han J., Lu J.P. Gas molecule adsorption in carbon nanotubes and nanotube bundles. Nanotechnology. 2002; 13:195–200.

[59] Lange U., Hirsch T., Mirsky V., Wolfbeis O. Hydrogen sensor based on a graphene–palladium nanocomposite. Electrochim. Acta. 2011; 56:3707.

[60] Hazra S.K., Basu S. Chemical modification of graphene and applications for chemical sensors. In: Thakur V.K., Thakur M.K., editors. Chemical Functionalization of Carbon Nanomaterials-Chemistry and Applications. 2015; CRC Press (Taylor and Francis), USA.

Section 4

Electrochemical Sensor

Chapter 7

Advances in Electrochemical Nitric Oxide Exhaust Gas Sensors

Erica Perry Murray and Ling Cui

Additional information is available at the end of the chapter

http://dx.doi.org/10.5772/62625

Abstract

The role of porosity on impedancemetric NO_x sensing is discussed for sensors composed of a porous yttria-stabilized zirconia (YSZ) electrolyte and dense Au electrodes. NO_x sensors considered here were fabricated at firing temperatures of 950–1200°C, which established a range of electrolyte microstructures where the porosity ranged from approximately 50% to 44%. Analysis of the electrical response of the NO_x sensors indicated that sensors fired at 1050°C resulting in an electrolyte porosity of 46% demonstrated higher NO_x sensitivity based on the operating conditions studied. The impedance of the sensors demonstrated a strong dependence on the electrolyte porosity. The activation energy of the sensors, which ranged from 109.2 to 81.1 kJ/mol, decreased with decreasing electrolyte porosity. Sensors with an electrolyte porosity ≥46% were limited by dissociated adsorption, whereas gas diffusion was rate limiting for sensors with an electrolyte porosity <46%. The impedancemetric response of the porous sensors to NO concentrations ≤10 ppm was distinguishable at operating frequencies as high as 40 Hz, thereby suggesting rapid sensing capabilities. Overall, the microstructure of the sensors composed of a YSZ electrolyte with 46% porosity promoted a strong, rapid, and highly sensitive response to NO_x.

Keywords: NO_x sensors, porous zirconia, impedancemetric, impedance spectroscopy, dense Au electrodes

1. Introduction

Automotive exhaust gas sensors play a critical role in monitoring emission components, such as nitric oxides, carbon monoxide, and hydrocarbons. These sensors also communicate with the vehicle on-board diagnostic system in order to regulate engine operation. Environmental regulations have largely been responsible for driving improvements in gasoline and diesel engine operation and efficiency, as well as the use of cleaner fuels, which have translated into lower pollutant levels [1–3]. Monitoring lower pollutant levels requires the application of sensors capable of detecting exhaust gas constituents with greater sensitivity, selectivity, and accuracy.

Nitric oxides (NO and NO_2), referred to as NO_x, are highly regulated exhaust gas species as these emissions significantly contribute to environmental and respiratory health issues. Vehicles with diesel engines tend to emit higher concentrations of NO_x emissions, in comparison to vehicles using gasoline engines. Thus, NO_x sensors are a crucial component in diesel exhaust aftertreatment systems. Ceramic oxide-based electrochemical sensors are widely used as NO_x sensors as they can rapidly respond to changes in the exhaust gas composition. Traditionally, these sensors have utilized a dense zirconia-based electrolyte accompanied by porous noble metal electrodes. The porous electrodes allow the exhaust gas to diffuse to reaction sites along the electrolyte/electrode interface within the sensor. This interface is called the triple-phase-boundary (TPB). As the exhaust gas diffuses, it is possible for heterogeneous catalysis reactions involving NO_x to occur within the electrodes, thereby, resulting in a lower concentration of NO_x at the TPB. This can be particularly problematic for accurately detecting low concentrations of NO and NO_2 gases. Current NO_x sensors can accurately detect NO_x concentrations down to approximately 10 parts per million (ppm). Martin et al and Ling et al [4, 5] have found that greater sensitivity can be achieved by changing the microstructure of sensor components, such that the electrolyte is porous and the electrodes are dense. Using this sensor architecture, which is opposite to conventional NO_x sensors, has enabled NO and NO_2 to be detected at concentrations as low as 5 ppm. The dense electrodes seem to limit heterogeneous catalysis reactions, while the porous electrolyte allows NO_x gases to proceed to the TPB where NO_x sensing reactions occur. The degree of porosity within the electrolyte, as well as the morphology, significantly influences the exhaust gas diffusion path, diffusion rate, and concentration of reaction sites at the TPB. Thus, in order to gain greater understanding of porous electrolyte properties that influence the electrochemical response of NO_x sensors it is necessary to understand the behavior of a broad range of electrolyte microstructures.

This chapter concentrates on the behavior of NO_x sensors composed of a porous yttria-stabilized zirconia (YSZ) electrolyte. YSZ is favored as the electrolyte in NO_x sensors due to its tolerance for both lean and rich exhaust gas conditions [6]. Under lean exhaust conditions, the oxygen concentration in the exhaust is high, whereas rich conditions create an oxygen-reducing environment. The yttria content contributes to the mechanical and ionic properties of the material. Fully-stabilized zirconia (i.e., 8 mol% Y_2O_3-stabilized ZrO_2) has a reasonably high ionic conductivity, which contributes to the magnitude of the NO_x sensitivity response [7]. The pores of the YSZ electrolyte provide pathways for gas diffusion, while the contiguous network of YSZ particles provides pathways for ionic transport. These pathways are dependent upon the electrolyte microstructure.

The various electrolyte microstructures presented in this chapter were studied using the impedancemetric method for NO_x sensor operation. Impedancemetric NO_x sensing is a relatively new approach that is favored for the potential to overcome performance limitations associated with other methods; and, lower manufacturing costs are expected to be more readily achieved by using this approach [4]. The following discussion will address the role of the NO_x sensor electrolyte microstructure, in terms of porosity, particle size, particle-to-particle contact, and TPB reactions, based on analysis of impedancemetric NO_x sensing for various operating conditions.

2. Impedance theory

Electrochemical impedance spectroscopy is a valuable and commonly used technique for examining the dynamics of mobile and bound charges within the bulk or at the interface of solid and liquid materials [8]. It is useful for investigating the electrochemical behavior of NO_x sensors, and can provide an understanding of the dependent relationship between NO_x reactions, and the composition, microstructure, and configuration of materials composing the sensor. Analysis of impedance data can also indicate potential reaction steps, such as adsorption, dissociation, diffusion, charge transfer, and ionic transport that can promote or limit the electrochemical behavior of the sensor. The impedance, Z, of a material is obtained by applying a small alternating voltage or current to the material sample and measuring the electrical response with respect to frequency, f. The impedance describes the opposition to current flow not only due to resistance, but also from capacitance and inductance effects that vary with frequency. The impedance is generally described in the complex plane by the real component, $Re(Z)$, and the imaginary component, $Im(Z)$, according to the following equations:

$$Z(\omega) = Re(Z) + jIm(Z) \tag{1}$$

where $\omega = 2\pi f$

$$Re(Z) = |Z| cos\theta \tag{2}$$

$$Im(Z) = |Z| sin\theta \tag{3}$$

The magnitude $|Z|$ is:

$$|Z| = \sqrt{[Re(Z)]^2 + [Im(Z)]^2} \tag{4}$$

and the phase angle, θ, is given by,

$$\theta = \tan^{-1}\frac{Im(Z)}{Re(Z)} \tag{5}$$

$|Z|$ or θ is commonly used to quantify the NO_x sensing response by the impedancemetric method. Some studies have found the θ response yields more sensitive and stable data in comparison to the $|Z|$ response [4]. For both $|Z|$ and θ, the sensing signal is frequency dependent, such that it is necessary to determine an appropriate frequency that enables noise factors to be minimized and the sampling rate to be maximized. Such conditions contribute to sensor accuracy, as well as a rapid response.

3. Experimental

Standard ceramic processing methods were used to fabricate NO_x sensors with a range of porous electrolyte microstructures. Tape cast 8 mol% Y_2O_3-stabilized ZrO_2 (YSZ, Tosoh Corp.) was used for the electrolyte along with gold (Au) wires that served as dense electrodes. The tapes were made by mixing YSZ powder with ethanol, binder (B-98 Butvar), dispersant (phosphate ester), and plasticizer (dipropylene glycol benzoate). The mixture was ball milled, dried, tape cast, and subsequently cut into rectangular strips that were approximately 10 mm × 5 mm × 1.5 mm. Further details concerning the fabrication steps are described elsewhere [5]. A YSZ slurry was made by dissolving additional YSZ tape into ethanol. This slurry was used to adhere layers of tape and coat the Au wire electrodes composing the NO_x sensors. One of the Au wire electrodes was wrapped outside of the assembly and the other Au wire was embedded between the YSZ tape layers. Additional YSZ slurry was coated over the entire sensor assembly. Illustrations of similar sensors were published by Cui and Murray [9]. The sensors were fired in air at 950, 1000, 1050, 1100, 1150, and 1200°C for 1 hour. The different firing temperatures resulted in different porous electrolyte microstructures and morphologies. Scanning electron microscopy (SEM) combined with statistical calculations and Archimedes method were used to analyze the porosity of the electrolytes. The statistical approach and calculations are discussed in reference [5].

Impedancemetric operation of the NO_x sensors was carried out using a Gamry Reference 600 that was set to apply a small alternating voltage of 50 mV across the sensor and record the electrochemical response over frequencies ranging from 1 Hz to 1 MHz. Experiments took place in a quartz tube within a furnace operated at temperatures of 600–700°C. A standard gas handling system composed of mass flow controllers was used to manage the supply of exhaust gas mixtures containing O_2, N_2, NO, and NO_2 for a total flow rate of 100 sccm. This flow rate was chosen to allow sufficient time for the gases to adsorb and diffuse through the pores within the electrolyte to the electrolyte/electrode interface. NO and NO_2 concentrations up to 100 ppm were studied with particular focus at lower concentrations (i.e., <20 ppm). These measurements were collected in the presence of 1% to 18% O_2 with N_2 as the balance gas. The phase angle component of the impedance, θ, was used to determine the sensitivity to NO and NO_2. Equivalent circuit analysis was used to model the impedance data in order to interpret activation energies, oxygen dependence, and potential rate limiting mechanisms impacting the NO_x sensing response.

4. Electrolyte microstructure and morphology

Typical SEM images of the various porous YSZ electrolytes fired at temperatures ranging from 950°C to 1200°C are shown in **Figure 1**. These images show the surface of the NO_x sensor electrolyte. The cross-sectional images (not shown) were very similar to the corresponding surface images. For sensors fired at lower temperatures, the YSZ particles were smaller in size and there was less contact between particles, in comparison to the electrolyte for sensors fired at higher temperatures. The SEM images showed YSZ particles of similar size, ranging from approximately 60 to 80 nm, for firing temperatures up to 1050°C. At 1100°C, there was a noticeable change in the microstructure as greater necking was observed between connected particles. As the firing temperature increased, the porous pathways and network of electrolyte particles became more distinct. Surface cracks were observed on electrolytes fired at 1150 and 1200°C. These defects were more substantial in electrolytes fired at 1200°C, which may indicate that thermal stresses were the source.

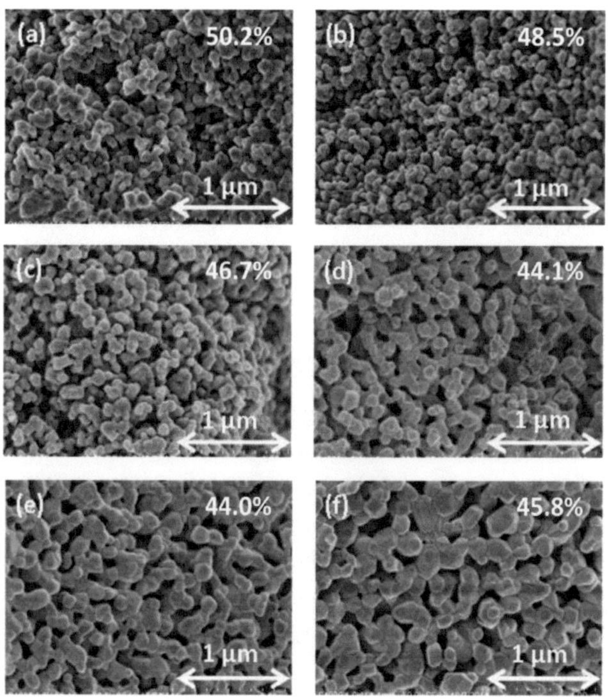

Figure 1. SEM images of YSZ electrolytes fabricated at (a) 950°C, (b) 1000°C, (c) 1050°C, (d) 1100°C, (e) 1150°C, and (f) 1200°C.

The porosity was calculated using a three-dimensional statistical method based on the difference in contrast between the pores and particles shown in the SEM images of the YSZ electrolytes. **Table 1** shows the porosity values determined for each sensor. The porosity decreased by about 6% for sensors fired between 950 and 1100°C. However, only a small change in porosity was observed for sensors fired at 1100 and 1150°C; and, an unexpected increase in electrolyte porosity was determined for sensors fired at 1200°C. It is likely that the cracks observed in the sensors fired at 1150 and 1200°C were a source of error in this data. Porosity measurements by Archimedes method, also shown in **Table 1**, were found to decrease with increasing firing temperature, and were consistently about 6% higher than the porosity calculations by the statistical method. Using Archimedes method, open pores composing the entire YSZ electrolyte were accounted for in the measurement; whereas the statistical method was based on a series of surface images of the electrolyte. Although Archimedes method is more prone to human error, it is possible that more reliable data, particularly for the electrolytes with surface cracks, are depicted in the results from Archimedes method.

T_f (°C)	Mathematical method		Archimedes method	
	Porosity (%)	σ (%)	Porosity (%)	σ (%)
950	50.16	1.82	56.6	0.87
1000	48.54	2.21	55.5	0.85
1050	46.70	1.88	52.3	1.11
1100	44.10	0.87	51.1	1.15
1150	44.06	0.66	50.6	0.64
1200	45.80	0.78	43.9	0.62

Table 1. Porosity results from mathematical method and Archimedes method as a function of the YSZ electrolyte fabrication temperature, T_f.

5. Impedancemetric NO_x response

The impedancemetric response of the various sensors to NO_x indicated similar data for the response to NO and NO_2. This is a common observation that has been associated with thermodynamic conversion of NO_2 to NO at elevated temperatures [5, 10]. Given this observation, representative data collected for NO are presented. Typical impedance data collected for the sensors indicated a strong relationship between the sensor electrolyte fabrication temperature and the magnitude of the impedance response. This can be seen in the impedance data collected for sensors with electrolytes fired at 1000, 1100, and 1200°C that were operated at 650°C in the presence of 10.5% O_2, shown in **Figure 2**. The data were measured in triplicate for each testing condition to insure that stable, reproducible data were collected. The sensor impedance response depends upon material composition and microstructure, molecular and electronic transport, as well as the following electrochemical reduction/oxidation reactions:

http://dx.doi.org/10.5772/62625

$$\frac{1}{2}O_2 + 2e^- \leftrightarrow O^{-2} \tag{6}$$

$$NO_2 + 2e^- \leftrightarrow NO + O^{-2} \tag{7}$$

The size of the electrolyte particles and degree of particle-to-particle connectivity influence the microstructure of the porous electrolyte, which affects gas and ionic transport pathways within the sensors. If gas transport through the porous microstructure or ionic transport through the particles is restricted, then reactions 6 and 7 will be impeded or limited. As shown in **Figure 1**, the electrolyte particle size and particle-to-particle connectivity increased with firing temperature, while the corresponding porosity decreased. Thus, it is necessary to establish a proper balance between gas and ionic transport pathways in order to promote NO_x sensitivity. The impedance data in **Figure 2** indicated a lower impedance resulted for sensors prepared at 1100°C, in comparison to the sensors prepared at 1000 and 1200°C. This result likely indicates that the microstructure of the sensors prepared at 1000°C lacked sufficient particle-to-particle connectivity, thereby limiting ionic transport and causing a larger impedance to be measured. On the other hand, the sensors prepared at 1200°C had more substantial particle-to-particle connectivity, but a lower porosity, which likely limited gas transport resulting in the higher impedance measurement. These data demonstrate the importance of understanding the behavior of gas and ionic transport pathways and their influence on optimizing sensor performance.

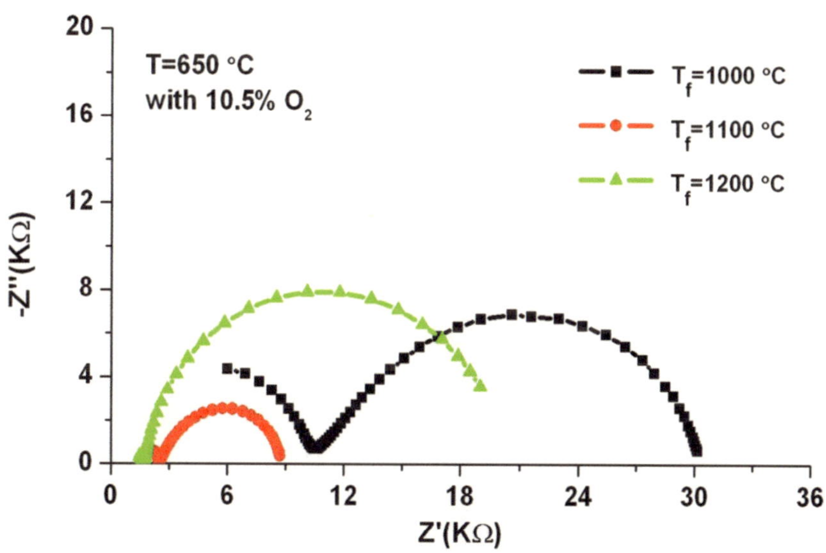

Figure 2. Nyquist plots of the sensors fired at 1000, 1100, and 1200°C while operating in 10.5% O_2 with N_2 as the balance.

To further analyze the electrical response of the sensors, equivalent circuit modeling with the Gamry EIS300 software was used to simulate the impedance measurements under various operating conditions. Bulk electrolyte properties associated with the porous YSZ microstructure, such as oxygen ion conductivity are described by the high-frequency arc (HFA) response. More importantly, the impedancemetric NO_x sensor response is primarily described by electrode reactions, which are presented by the low-frequency arc (LFA) response. The equivalent circuit model that best fits the sensor impedance results was (R_1CPE_1)(R_2CPE_2), as shown in **Figure 3**.

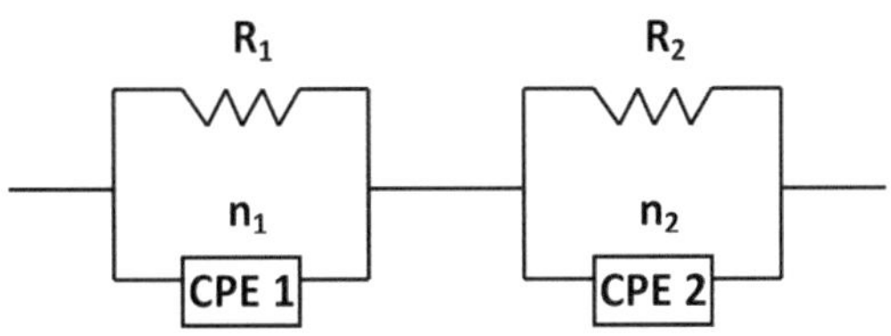

Figure 3. Equivalent circuit used to model the impedance data.

The resistor, R_1, was associated with the HFA and interpreted as the ionic transport resistance of the porous YSZ electrolyte. The HFA depends upon the porosity of the electrolyte as the porous microstructure creates tortuous pathways that can impede the flow of oxygen ions. The LFA resistance, R_2, described the interfacial resistance associated with TPB reactions. Such reactions include gas adsorption, dissociation, diffusion, and charge transfer. The equivalent circuit model accounted for the slight suppression of the arcs with constant phase elements, CPE_1 and CPE_2, which described the nonideal capacitance behavior of the electrolyte and electrodes, respectively. The CPE is defined by the following impedance equation:

$$Z(\omega) = \frac{1}{Y_0 (j\omega)^n} \tag{8}$$

where the impedance, Z, depends on the angular frequency, ω, and Y_0 is a constant. When $n = 1$, the $Z(\omega)$ represents a capacitance with a value equivalent to Y_0. Surface defects, interfacial reactions, as well as the morphology of the sensor components can cause the time constant for various reactions to differ resulting in nonideal capacitance behavior [8]. **Table 2** shows the simulated data generated for the sensors. These data were used to calculate the capacitance, C, associated with the high and low frequency sensor response according to the following equation:

$$C = \frac{\left[(R)(\mathrm{CPE})\right]^{\frac{1}{n}}}{R} \tag{9}$$

The CPE represents an ideal capacitor when $n = 1$. The n values shown in **Table 2** were relatively high, which indicated that the deviation from ideal capacitance behavior was small. Thus, the capacitance values determined from Eq (9) were reasonably accurate. The capacitance is associated with oxygen coverage at the Au/YSZ interface [9, 10]. The data in **Table 2** also show that the capacitance tended to decrease when NO was added to the test gas for sensors fabricated at temperatures up to 1100°C, which suggests that oxygen coverage diminished as NO occupied more sites along the TPB. The slight increase in capacitance observed in sensors prepared at 1150 and 1200°C may be related to the surface defects observed in these samples. Further discussion of the influence of oxygen on sensor behavior will be addressed in the later section of this chapter.

T_f (°C)	High-frequency arc		Low-frequency arc			
	10.5% O_2		10.5% O_2		10.5% O_2 + 100 ppm NO	
	Y1(F) 10^{-10}	n_1	Y2(F) 10^{-6}	n_2	Y2(F) 10^{-6}	n_2
950	0.638	0.89	0.427	0.77	0.416	0.77
1000	0.133	1.00	0.829	0.68	0.638	0.73
1050	1.149	0.86	0.923	0.79	0.909	0.79
1100	2.727	0.82	2.912	0.82	2.799	0.83
1150	2.550	0.83	1.867	0.92	1.871	0.92
1200	0.265	1.00	2.162	0.91	2.260	0.90

Table 2. Equivalent circuit fitting parameters are shown for sensor impedance data collected at 650°C.

NO_x sensitivity based on the impedancemetric method can be assessed using various parameters, including the angular phase response, magnitude of the impedance, and capacitance behavior. Here the angular phase response, θ, at specific frequencies, f, was used to measure NO sensitivity for the sensor with various porous electrolytes. The angular phase response is favored for yielding highly accurate data with limited noise contributions. An operating frequency of 40 Hz was determined from the impedance data to be a suitable frequency for monitoring the angular phase response of the sensors for various concentrations of NO, in order to achieve a relatively high sampling rate and response time. Based on the angular phase response, the sensitivity, $\Delta\theta/\theta$, of the sensors to NO was determined using the following equation: $\Delta\theta/\theta = [(\theta_{O2} - \theta_{NO})/\theta_{O2}]$. The term θ_{O2} corresponds to the baseline phase angle response when only O_2 was present, and θ_{NO} represents the phase angle response for a specific NO concentration. **Figure 4** shows the NO sensitivity gas experiments for the porous electrolyte sensors for operation at 650°C as the NO gas concentration was varied. The gas sensitivity data indicate that sensors prepared at 1050°C with an electrolyte porosity of 46% demonstrated greater NO sensitivity, particularly at lower concentrations of NO, in comparison to sensors fabricated at different temperatures. This suggests that the microstructure achieved at 1050°C promoted sufficient gas and ionic transport to enable reactions contributing to NO sensitivity to proceed more readily. It is important to note that although the gas experiments also indicated reasonably high NO sensitivity responses by sensors fabricated at 950 and 1000°C, the higher firing temperature of 1050°C results in a stronger and more durable ceramic electrolyte. Thus, the microstructural and mechanical properties of electrolytes prepared at 1050°C contributed to greater gas sensitivity to NO along with greater sensor durability.

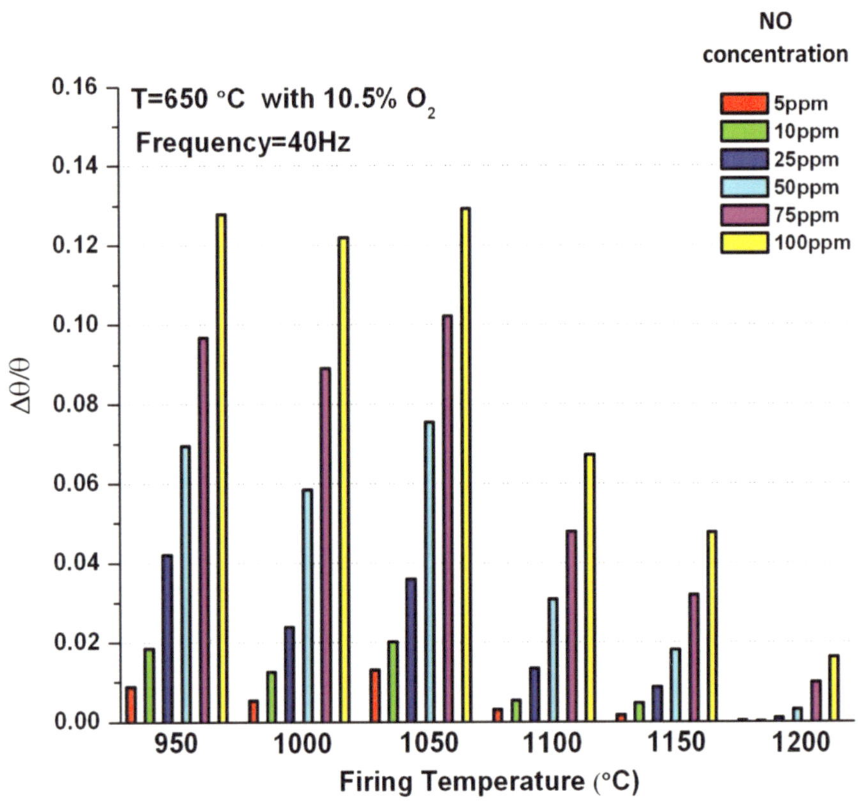

Figure 4. The fractional phase angle response for sensors fired at 950–1200°C in the presence of NO at a frequency of 40 Hz.

Knowledge of the oxygen and temperature dependence of the sensors can greatly aid interpretation of mechanisms and reaction kinetics governing NO sensing behavior. The oxygen partial pressure dependence was determined from the low-frequency arc, (R_1CPE_1) impedance data collected for each sensor. As shown in **Figure 5**, a power law dependence of $R_1 \propto (pO_2)^m$, where the slope m ≈ −0.47 to −0.72 was observed for the various porous electrolyte sensors. The slope values were found to become more negative as the microstructure of the electrolyte became less porous. Studies concerning oxygen partial pressure dependence have reported surface limiting reactions, such as dissociative adsorption or atomic oxygen as potential rate limiting mechanisms when the slope, m = −0.5 [11, 12]. Limited gas diffusion is commonly associated with slope values approaching −1. The observations of the data shown in **Figure 5** support gas diffusion becoming a rate limiting mechanism as the electrolyte porosity decreased and the gas transport channels became more limited in the sensors prepared at higher firing temperatures. The sensors fabricated at 950–1050°C appeared to be limited by surface limiting processes. Since greater NO sensitivity was achieved for these sensors, it seems plausible that surface limiting processes do not significantly limit NO sensing behavior under the operating conditions tested.

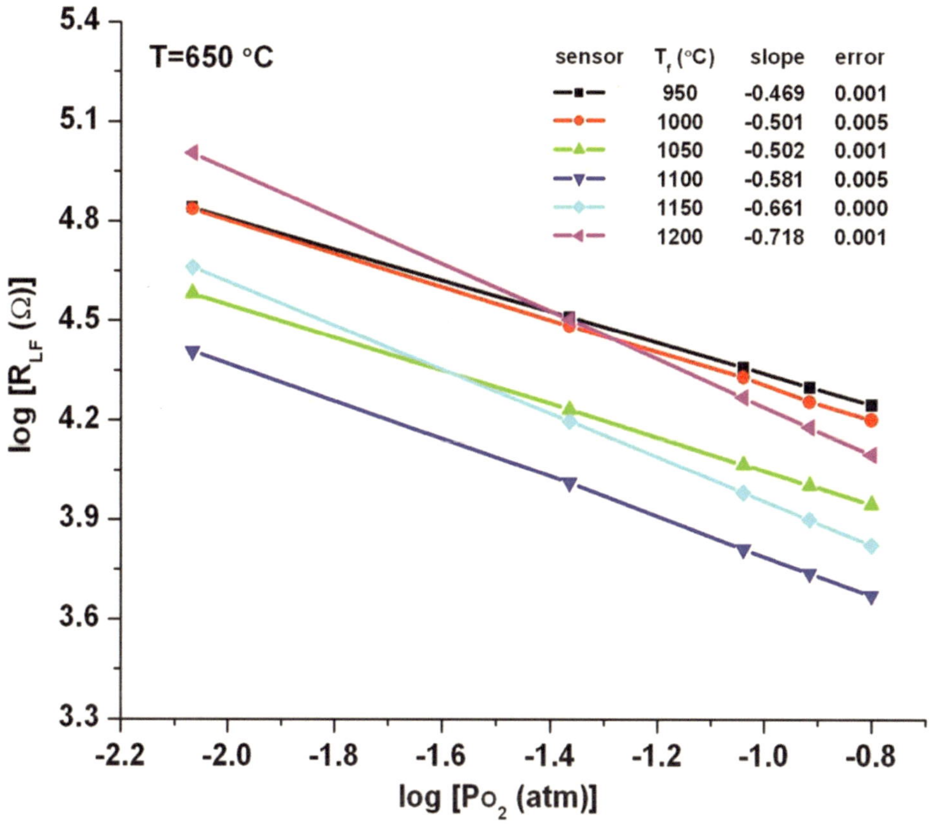

Figure 5. Oxygen partial pressure dependence of the various sensors at an operating temperature of 650°C under conditions where the oxygen concentration was varied from 1% to 18%.

The temperature dependence of the sensors was determined by the Arrhenius plots of the data that primarily described electronic/ionic transport behavior. From the slope of the plotted data, the activation energy, Ea, was determined for each sensor. The data in **Figure 6** indicated that oxygen ions were able to travel more readily as the operating temperature increased; and, that greater ionic transport was achieved as the porosity of the sensor electrolyte decreased. Sensors with lower porosity had larger YSZ particles and greater particle-to-particle connectivity, which supported greater ionic transport. The activation energy values presented here were typical for porous YSZ electrolyte NO_x sensors with Au electrodes. The activation energy data in **Figure 6**, taken into consideration with the NO sensitivity data in **Figure 4**, suggest that sufficiently high oxygen ion transport activity can limit NO sensitivity. Thus, a microstructure that creates a suitable balance between the gas and ionic transport pathways is necessary for establishing optimal NO_x sensing properties.

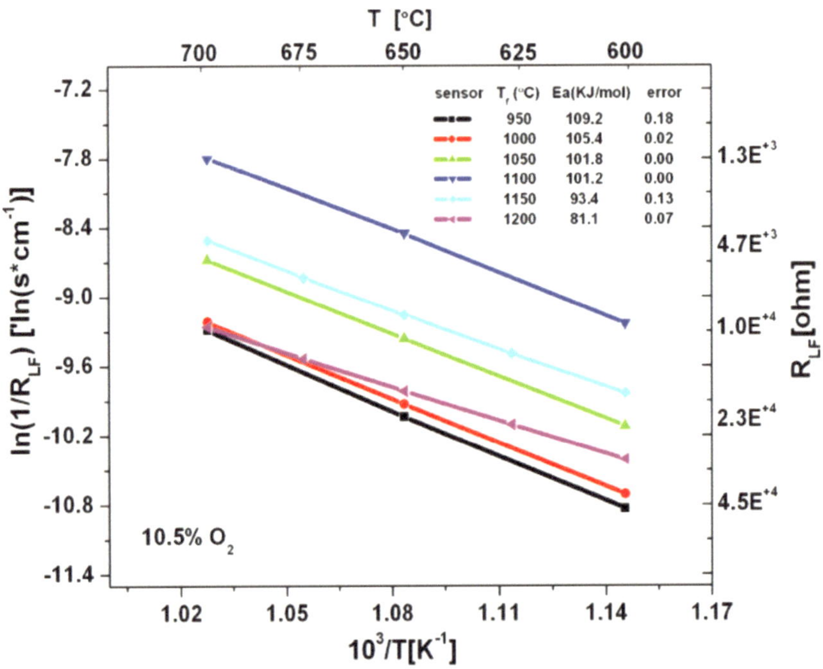

Figure 6. The temperature dependence is described by the Arrhenius plot for the sensors fired at 950–1200°C along with corresponding activation energies.

6. Conclusions

Porous electrolyte-based NO_x exhaust gas sensors are a novel and promising concept for expanding the capabilities of automotive gas sensing technology. The work presented here using the impedancemetric method clearly demonstrated the influence of the porous electrolyte microstructure on NO_x sensing behavior. Substantial differences in the impedance response, sensitivity, rate limiting mechanisms, and activation energies were observed in the data collected for NO_x sensors with porous electrolytes ranging from 50% to 44%. The increase in NO_x sensitivity observed in the sensors fabricated at 1050°C with a corresponding electrolyte porosity of 46% was promoted by establishing a microstructure that provided a proper balance between the gas transport pathways created by the porous microstructure and the ionic transport pathways formed by the YSZ particles. In addition, the resulting particle size and particle-to-particle contact of sensors with an electrolyte fabricated at 1050°C formed an electrolyte/electrode/gas interface (i.e., TPB) that enabled reactions favoring NO_x to proceed more readily. The phase angle measurements collected at 40 Hz indicated NO_x sensitivity can be achieved at significantly low concentrations (e.g., 5 ppm), which suggests that rapid sensing

http://dx.doi.org/10.5772/62625

is feasible for such sensors. Within the range of porosities tested, surface limiting reactions appeared to be a rate limiting mechanism based on the oxygen partial pressure dependence. However, this limiting mechanism did not appear to significantly impede NO_x sensing capabilities for sensors prepared at 1050°C. A strong relationship was observed between the sensor electrolyte porosity and the sensor activation energy. Lower activation energies correlated with lower electrolyte porosity. As research continues, the outcomes are expected to contribute to a more complete understanding of NO_x sensor behavior and provide fundamental knowledge necessary for the advancement of NO_x exhaust gas sensor technology.

Acknowledgements

The authors thank Dr. Weizhong Dai and Dr. Fei Han for carrying out the mathematical calculations used to determine the porosity of the electrolyte-based NO_x sensors. The authors also extend their appreciation to Mr. Robert Novak and Dr. Jaco Visser of Ford Motor Company for providing technical support and meaningful discussions that contributed to this work. Funding of this work was provided by the National Science Foundation under the Ceramics Program (DMR-1410670).

Author details

Erica Perry Murray* and Ling Cui

*Address all correspondence to: emurray@latech.edu

Institute for Micromanufacturing, Louisiana Tech University, Ruston

References

[1] Bandivadekar A, Bodek K, Cheah L, Evans C, Groode T, Heywood H, Kasseris E, Kromer M, Weiss M. On the Road in 2035: Reducing Transportation's Petroleum Consumption and GHG Emissions. MIT Laboratory for Energy and the Environment, Cambridge, MA, Report No. LFEE 2008-5 RP, July 2008.

[2] United States Environmental Protection Agency, 2007 Progress Report: Vehicle and Engine Compliance Activities, EPA-420-R-10-022, Aug 2010.

[3] United States Environmental Protection Agency, EPA's Program for Cleaner Vehicles and Cleaner Gasoline, EPA420-F-99-051, Dec 1999.

[4] Martin L, Woo L, Glass R. Impedancemetric NO_x sensing using YSZ electrolyte and YSZ/Cr_2O_3 composite electrodes. Journal of the Electrochemical Society. 2007; 154(3):J97-J104, doi:10.1149/1.2430646.

[5] Ling C, Fei H, Weizhong D, Erica PM. Influence of microstructure on the sensing behavior of NO_x exhaust gas sensors. Journal of the Electrochemical Society. 2014; 161(3):B34-B38, doi:10.1149/2.088403jes.

[6] Jan VH, Augustin JM, Thampi KR. Conductivity measurements of various Yttria-stabilized zirconia samples. Journal of Materials Science. 1994; 29(14):3691–3701, doi: 10.1007/BF00357336.

[7] Praveen KS, Eric LB, Rangachary M, Mark AN, Dennis T, Fernando HG. Effect of yttria-stabilized zirconia sintering temperature on mixed potential sensor performance. Solid State Ionics. 2010; 181(19):947–953.

[8] Evgenij B, Ross M. Impedance spectroscopy theory, experiment, and applications. New Jersey: John Wiley & Sons, Inc., 2005.

[9] Ling C, Erica PM. Effect of electrode configuration on nitric oxide gas sensor behavior. Sensors. 2015; 15:24573–24584. doi: 10.3390/s150924573.

[10] Woo L, Martin L, Robert SG, Wensheng W, Sukwon J, Raymond JG, Erica PM, Rober FN, Jaco HV. Effect of electrode composition and microstructure on impedance metric nitric oxide sensors based on YSZ electrolyte. Journal of the Electrochemical Society. 2008; 155(1):J32-J40, doi:10.1149/1.2804766.

[11] Takeda Y, Kanno R, Noda M, Tomida Y, Yamamoto O. Cathodic Polarization phenomena of perovoskite oxide electrodes with stabilized zirconia. Journal of the Electrochemical Society. 1987; 134(11):2656–2661, doi:10.1149/1.2100267.

[12] Fukunaga H, Koyama M, Takahashi N, Wen C, Yamada K. Reaction model of dense $Sm_{0.5}Sr_{0.5}CoO_3$ as SOFC cathode. Solid State Ionics. 2000; 132:279–285.

Section 5

Acoustic Chemical Sensor

Chapter 8

Sensing Materials for Surface Acoustic Wave Chemical Sensors

Mohd Nizar Hamidon and Zainab Yunusa

Additional information is available at the end of the chapter

http://dx.doi.org/10.5772/63287

Abstract

Online real-time monitoring of gases requires a miniaturized, passive, and accurate gas sensor. Surface acoustic wave (SAW) devices possess these properties which make them suitable for gas-sensing applications. They have shown remarkable results in sensing of different gases in terms of sensitivity, selectivity, response, and recovery times. One of the important prerequisites a designer should know is to have knowledge on the different types of sensing material suitable for gas-sensing applications, prior to design and fabrication of the sensor. Different sensing materials, including metal oxides, polymers, carbon nanotubes, graphene, nanocomposites, etc. have been used for SAW gas sensors. In this article, different sensing materials for SAW gas sensors will be discussed.

Keywords: surface acoustic wave, gas sensor, carbon-based material, polymer, resonator

1. Introduction

Acoustic wave technology can offer the technology for gas detection and has been used for a variety of wireless sensor applications for some decades. The Surface Acoustic wave (SAW) technology has offered the development of small, lightweight, battery-free, maintenance-free, and multiple-sensor wireless interrogation operations [1]. SAW components have been used as filters or resonators in mobile phones in the telecommunication industry. They are also used as sensors for pressure, torque, acceleration, humidity, temperature, chemical, and biological applications.With the advent of modern technology, lots of efforts have been made by researchers to reduce air pollution by the development of highly responsive gas sensors. Different SAW configurations which include delay lines and resonators have proved to be reliable and showed good sensitivity, selectivity, and response time. The advantage of their small size and having tunable frequency has made them rather promising sensing devices.

In order to develop a highly sensitive sensor, knowledge for the selection of suitable sensing materials becomes rather critical. In this article, an insight into different sensing materials for SAW gas sensors is presented.

1.1. Sensing materials

As the gas sensor consists of a transducer and a sensing layer, the sensing layer therefore constitutes an important part of the SAW sensor. The mechanism of sensing is such that any changes in the conductivity, mass loading, viscoelastic effect, stress effect, as well as charge carriers from gas adsorption of the sensing material cause a corresponding frequency shift. This relationship had been represented by Sauerbrey [2]. It shows the relationship between the mass that has been deposited on the SAW device and the resonance frequency. Based on this equation, the density, shear modulus, loaded mass, and the type of piezoelectric substrate, all contribute to the sensing effect. When a sensing material is deposited on a SAW, there is viscoelastic effect on the SAW propagation, which leads to a shift in acoustic wave velocity and also changes in acoustic wave attenuation. The shift in wave velocity is as a result of mass loading, while the acoustic attenuation is led by the viscoelastic properties of the sensing material. Moreover, dynamic behavior significantly influences the response of the device. In case of polymers, there is an assumption that the device responds to the mass of the polymer film. This assumption is valid only if the film is rigid and moves synchronously with the surface of the resonator that is oscillating. As a result of the moving film, the kinetic energy produced causes a decrease in the resonant frequency. However, if the moving film does not move synchronously with the surface of the film, it causes a lag behind the resonator, and this creates a viscoelastic response.

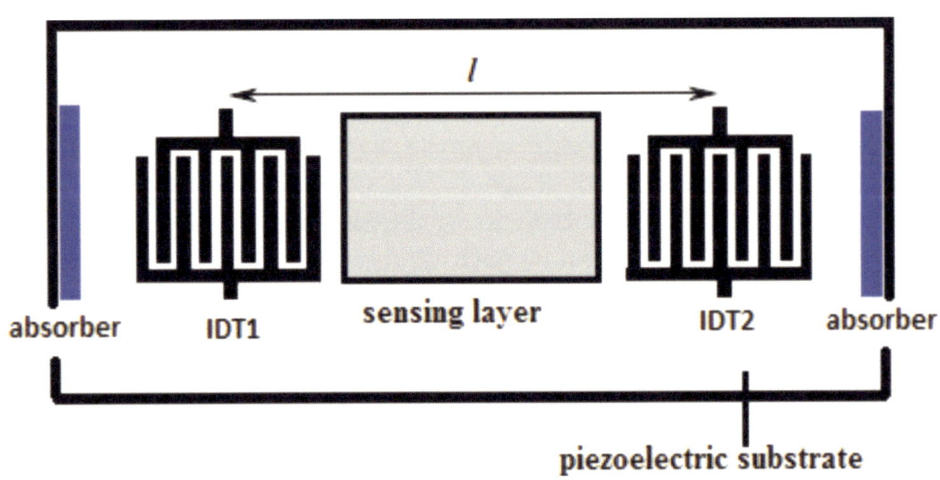

Figure 1. Two-port SAW delay line gas sensor.

The sensing layers are responsible for interacting with the different types of target gases. **Figure 1** depicts a SAW-based gas sensor which consists of the interdigital transducers (IDTs), reflectors, and the sensing layer. One-port and two-port SAW delay lines show the interdigital transducers (IDTs), reflectors, and the sensing layer. Surface acoustic waves are generated on a piezoelectric substrate with two sets of IDT. The first transducer converts electrical signal into acoustic wave, while the second converts this back to electrical signal. Rayleigh waves are usually used for gas-sensing applications [3]. The mechanism of SAW-based sensor is that when a thin sensing layer is applied on a piezoelectric substrate upon gas exposure, there is an interaction between the gas molecules and the sensing layer. This interaction will cause perturbations on the boundary conditions of the propagating SAW wave, which manifest as a change of velocity and attenuation of the propagating wave.

In order to choose a suitable sensing layer, several factors need to be considered. These factors include good selectivity toward the target gas, good response and recovery time, stability of the material over time, affordable cost, and nontoxicity of the material.

The sensitivity of the sensor is measured based on the output measurand of the sensor which could be voltage, current, resistance, or frequency. In the case of SAW sensor, the output is based on frequency; therefore, it is defined as $\Delta f/\Delta c$, where Δc is the change in analyte concentration. S is expressed in terms of Hz/ppm or Hz/vol. %. Similarly, in the case of resistive or conductometric sensor, the sensitivity is expressed as $\Delta R/R$ or $\Delta G/G$.

Therefore, knowledge of the different types of sensing material is essential for careful selection, so as to fulfill the above-mentioned characteristics.

There are different types of sensing materials used for gas-sensing applications. These include metal oxides, polymers, nanocomposites, carbon nanostructures which include nanotubes, nanofibers, graphene, and their composites, etc.

1.1.1. Metal oxides

The use of metal oxides as gas-sensing materials was first reported by Brattain and Berdeen in 1953 [4]. Since then, a lot of researches have been done toward investigation of different types of metal oxides for sensing applications. Metal oxides are materials which are widely used in different types of gas sensors.

The specific characteristic that makes them universal to be used in gas sensors includes high thermal stability which makes them suitable for applications in harsh environments. They also have good adaptability to different reducing or oxidizing gases by changing their electrical properties in the presence of gases. This conductivity change manifests as a change in resonance frequency, which is translated as the sensor signal.

Different types of metal oxides have been employed in SAW gas sensing for the detection of different gases. Metal oxides suitable for applications in SAW gas sensing include tin oxide (SnO_2), Indium oxide (In_2O_3), tungsten trioxide (WO_3), and zinc oxide (ZnO). These metal oxides have been employed as SAW sensing materials for the detection of different gases, including hydrogen, ammonia, carbon monoxide, ethanol, etc. **Table 1** summarizes the types of metal oxides used for SAW gas sensing.

Author	Type of metal oxide	Gases detected	Temperature
Wang et al. [6]	In_2O_3	Hydrogen	Room
V.B Raj et al. [7]	ZnO, SnO_2, TeO_2	Ammonia	–
Duy Chan et al. [8]	Pt/ZnO	Hydrogen	Room
A.Z Sadek et al. [9]	ZnO	Hydrogen	200–300°C
Tang et al. [10]	Co_3O_4/SiO_2	Ammonia	Room
Tang et al. [11]	ZnO/SiO_2	Ammonia	Room
	ZnO		200°C-300°C

Table 1. Summary of some metal oxides for gas-sensing application.

1.1.1.1. SAW hydrogen metal-oxide based gas sensor

The first hydrogen SAW sensor was reported by D. Amico in 1982 [5], whereby palladium was used for hydrogen sensing; however, for improved sensitivities, metal oxides used for hydrogen sensing include indium oxide, zinc oxide, tungsten trioxide, etc. Indium oxide is also one of the metal oxides that is used for sensing both reducing and oxidizing gases. It has been widely used for the detection of different gases with improved sensitivities developed a SAW hydrogen sensor based on indium oxide at room temperature [6]. The experimental results showed a high sensitivity for hydrogen gas with a frequency shift of 11.83 kHz toward 400 ppm of hydrogen gas.

Zinc oxide was among the earliest metal oxides discovered and is widely used due to its good thermal and chemical stability and high conductivity of mobile electrons. However, the alignment of well-ordered nanostructure improves the performance of the sensor. Sadek [12] developed a ZnO nanorod SAW gas sensor for hydrogen detection. Results obtained showed good sensing performance when exposed to different hydrogen concentrations at operating temperatures of 200–300°C. The highest frequency shift was obtained at an operating temperature of 265°C.

In order to enhance the performance characteristics for hydrogen sensing at room temperature, many noble metals including Pd, Pt, Au, Ti, and Ni are employed. Duy-ThachPhan et al. [8] fabricated a SAW hydrogen gas sensor by using ZnO nanoparticles incorporated with a Pt catalyst. The integration of Pt/ZnO formed a layered structure on the SAW device in order to improve the sensitivity. Based on [13], Pt is the best catalyst to dissociate hydrogen atoms at room temperature. Experimental results obtained showed good results with a frequency shift of 55 kHz in 1% hydrogen concentration at room temperature.

http://dx.doi.org/10.5772/63287

1.1.1.2. SAW ammonia-based metal oxide sensors

The first SAW ammonia gas sensor was reported in 1987 in which platinum film was employed as a sensing layer for ammonia detection [14]. Since then, different types of materials are explored for ammonia sensing. Metal oxides used for ammonia sensing include WO_3, ZnO, Cobalt III oxide, etc.

WO_3 is one of the n-type semiconductor materials which have been employed for gas-sensing applications. Their property of having large surface–volume ratio has made them suitable for this purpose. Past researches have showed good sensitivity toward NO_2, H_2S, O_3, and H_2 [15–17]. It has been used as a sensing layer for the detection of concentrations of less than 1% of hydrogen gas in air [18]. However, for improved sensitivity, the structure of the tungsten trioxide was modified by sputtering platinum (Pt) and gold (Au) metals, and the sensing behavior was investigated at various operating temperatures [19].

Recently, nanowires have also been used on WO_3 in order to enhance the sensitivity and recovery times of hydrogen at room temperature [20]. In order to improve the sensitivity and selectivity, oxides are mixed with CNT to form composite and used for ammonia sensing. A highly sensitive ammonia gas sensor was fabricated using SnO_2/MWCNTs composites at room temperature [21].

In order to improve the sensitivity of SAW devices, layered SAW devices have been fabricated. Results obtained have shown an improved sensitivity, when compared to nonlayered devices [22]. The layered structure (**Figure 2**) shows a configuration with a ZnO intermediate layer and also WO_3 sensing layer.

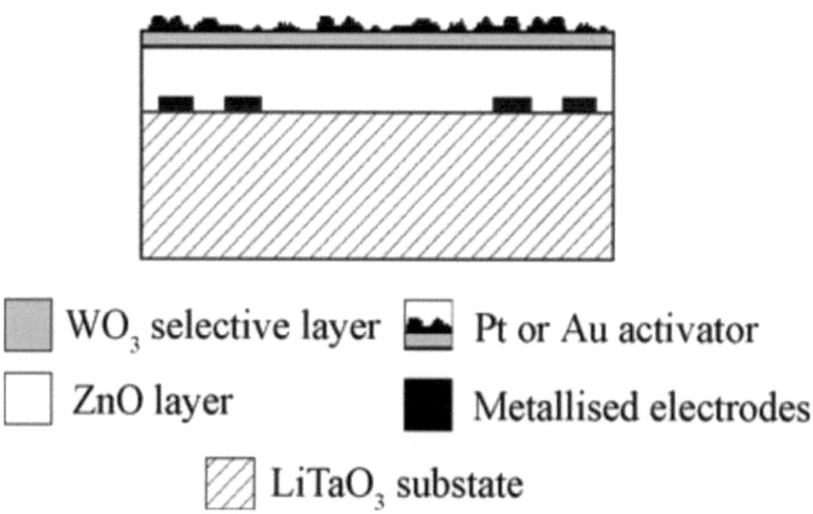

Figure 2. Layered SAW sensor [23].

The function of the intermediate layer is to increase the sensitivity arising from conductivity changes at the surface. It also traps the acoustic energy at the substrate. This configuration has been used for the detection of hydrogen [23, 24], ethanol vapor, and humidity [25].

Cobalt III oxide has also been considered as a p-type semiconductor, and it possesses the nonstoichiometric property. It has been used for sensing different gases, such as CO, $H_{2,}C_2H_5OH$, and NO_2 gases [4], but proved nonresponsive with ammonia gas. It also has the limitation of high operating temperatures above 200°C. Recently, Yong-Liang et al. fabricated a SAW-based ammonia gas sensor using hybrids of Co_3O_4/SiO_2[10]. Results obtained showed improved sensitivities when compared with pristine $Co_3O_4$and $SiO_{2.}$

Recently, different thin films of oxides were used in order to investigate the sensing behavior toward SAW-based ammonia gas sensor. Thin films of ZnO, SnO_2, TeO_2, and TiO_2 were deposited in the SAW sensor with the same thickness. Results obtained showed that the ZnO-coated SAW was found to be more sensitive toward liquor ammonia, compared to the other oxides [7].

1.1.2. Polymers as SAW sensing materials

Conductive polymers are another class of promising materials that are suitable for applications in gas sensing. The advantage of polymers over metal oxides is that they can work efficiently even at room temperature. Another advantage of conducting polymers is the ease of integration to the transducers of SAW device.

They comprise of carbon and hydrogen as the major constituents, while nitrogen, oxygen, sulfur, phosphorus, and halogens as minor constituents. The polymers that are promising for different sensing applications include polyacetylene, polythiophene, polypyrrole, polyaniline, etc. Polyaniline (Pani) and polypyrrole (PPy) are p-type semiconductors that are very unstable in the undoped state. These polymers have conjugated π-electronic system that changes electronically upon gas interaction.

When polymers are deposited to the SAW device and deployed for gas-sensing applications, the changes in the frequency are not due to the adsorption of gas/analyte, but as a result of the change in the modulus of the polymer [2] . The layers have an influence on the wave propagation velocity, which is dependent upon the viscoelastic effect of the polymers.

The viscoelastic responses of SAW could be in two forms. In the first mechanism, the response of the SAW is tracked by the velocity (v) changes related with the changes in storage modulus (G'), and the viscoelastic responses and attenuation (a) changes are related with the change in loss modulus (G''). Similarly, the second mechanism showed that in a thickness shear mode resonator, the movement of polymer films is not always synchronous with that of the surface of the device [26]. Rigid films with large value of G and phase lag move synchronously with the surface of the oscillating resonator, whereas compliant films with lower G and high tend to lag behind the resonator. Therefore, the phase shift is also one of the factors that leads to changes in the film dynamics and viscoelastic response. A model has been developed by Grate et al., in which the frequency shift was described as a resultant effect of vapor mass adsorption and the swelling of the polymer layer.

http://dx.doi.org/10.5772/63287

To improve the sensitivity, the surface of the SAW sensor can be modified using surface treatments so as to obtain selected chemical properties as reported by [27]. Polyisobutylene was employed as a sensing layer in a SAW sensor for the detection of perchloroethylene (PCE).

Ricco et al. [28] reported lead phthalocyanine (PbPc) conducting thin films employed as SAW sensing material for NO_2detection. Results obtained showed increase in sensitivity in the magnitude of 1000 times as compared with pure mass response of SAW devices. However, due to continuous exposure to gas, the sensor became oxidized, which leads to a drift in sensor signal. Humidity interference is one of the limitations of conducting polymers.

Polypyrrole nanofibers were used as a sensitive layer for hydrogen detection. They were synthesized using a template-free chemical process and deposited in the SAW sensor. Responses showed frequency shift of 20 kHz toward 1% of hydrogen gas and 4.5 kHz for 2.1 ppm of NO_2 [29]. However, for improved sensitivities, bilayer structures of metal-free phthalocyanine (H_2Pc) with palladium (Pd) have been employed for hydrogen gas sensing at a very low temperature. To reduce the effect of humidity on gas sensing, a layer of polyethylene membrane was placed on the bilayer structure. Frequency shifts were observed for hydrogen concentrations between 0.5 and 4% [30].

Polythiophene nanofibers have also been used for hydrogen detection. They have also been synthesized using the template-free method and deposited on a layered SAW surface. Frequency shift was observed to be 17 kHz toward 1% of hydrogen concentration [31].

Similarly, polyaniline nanofibers were employed as a sensing layer for the detection of hydrogen gas. A cross section of the SAW sensor is shown in **Figure 3**. The nanofibers have been synthesized using electropolymerization process and then subsequently deposited on the SAW substrate. The thickness of the dedoped polyaniline was varied, and the response behavior to hydrogen gas was investigated. Frequency shifts were observed for all thicknesses, with the smallest thickness having frequency shift of 12.1 Hz for 0.06% of hydrogen, while the largest thickness SAW sensor gave a frequency shift of 9.2 kHz [32].

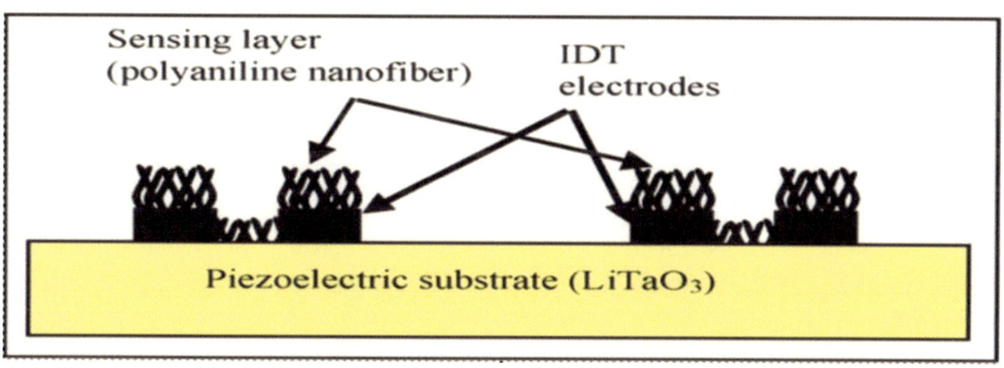

Figure 3. SAW gas sensor based on dedoped polyaniline nanofiber [32].

The integration of metal oxides with polymers has a great potential for increased sensitivity and selectivity at room temperature. As reported by [33], nanostructured polyaniline has the ability of high diffusion as a result of high surface-to-volume ratios and increased penetration depth for the gas molecules. Similarly, metal oxide nanoparticles are expected to improve sensor performance due to the grain size and high surface-to-volume ratios. Therefore, with the hybridization of polyaniline/SnO_2, hydrogen response is greatly enhanced with 30 s response time and 120 s recovery time to 1% hydrogen gas concentration at room temperature. The same authors [34] also reported the use of polyaniline/WO_3 sensing layer on a two-port SAW resonator for hydrogen detection at room temperature. Sensor performance was greatly improved with 40 s response time and 100 s recovery time to 1% hydrogen gas at room temperature.

Atashbar et al. developed a SAW-based sensor with camphorsulfonic acid (CSA) polyaniline nanofibers as a sensing layer for the detection of hydrogen at room temperature.

1.1.3. Carbon-based materials for gas-sensing applications

The advent of nanotechnology has attracted much interest into the use of carbon-based materials for gas-sensing applications. They possess inherent properties that make them attractive, which include high thermal and mechanical stabilities, high surface area, and good metallic and semiconducting behaviors. Carbon-based materials include carbon nanotubes (CNTs), nanofibers, nanobelts, nanorods, graphene, etc. The employment of CNTs for different gas-sensing applications is more common among researchers.

CNTs are of two types: single-walled carbon nanotubes (SWCNTs) and multi-walled carbon nanotubes (MWCNTs). A SWCNT is formed from a roll of graphene formed into a cylinder with diameters ranging from 0.4 to 2 nm, while MWCTs are formed from several layers of graphene rolled into nested rings of cylinder. They can be synthesized using three methods: laser ablation, arc-discharge, and chemical vapor deposition. Pristine CNTs possess very strong sp^2 carbon–carbon bonds, which gives them a very good stability. Their small diameter and hollow structure make them suitable for gas adsorption and detection. The small diameter allows electrical signals flowing along the tube to interact with any slightest defects within or outside the tubes. Both MWCNTs and SWCNTs are suitable for gas detection due to their semiconducting and metallic behaviors.

1.1.3.1. Gas sensors based on MWCNTs

MWCNTs have been used successfully for the detection of different gases, CO, NO_2, $H_{2,}NH_3$, and N_2[35]. Their drawbacks arelong recovery time as shown by and that sensing is required at high temperature [36–38]. Jayatissa [39] investigated the detection of hydrogen gas using MWCNT films synthesized by CCVD. Sensing of hydrogen gas was observed only at temperatures between 150 and 300°C, but there was no sensing at higher temperature (400°C) or low temperature (25°C). Similarly, MWCNT-based sensor was also employed for the detection of NO; to increase the sensitivity, an electric field was introduced between the two copper plates, with one of them containing the MWCNT silicon wafer. It was observed that the stronger the

http://dx.doi.org/10.5772/63287

electric field applied, the better the sensitivity would be [40]. Sayago [37] fabricated an NO_2gas sensor using double-walled and MWCNT, respectively. Sensing was observed at 25–250°C toward NO_2, but not H_2or NH_3. Good sensing response was only observed at temperatures higher than 100°C.

Similarly, Penza et al. [41] fabricated a SAW gas sensor based on SWCNT and MWCNT for the detection of volatile organic compounds. The CNTs were deposited on the SAW transducer using the spray painting technique, as shown in **Figure 4**. Results obtained showed high sensitivity toward ethanol, ethyl acetate, and toluene at room temperature. The authors observed and reported that the SWNT showed a higher sensitivity than the MWCNT solvent for the dispersion of the CNT, and also affects the sensing behavior. The sensing ability of SWNT is better than the MWCNT, because the SWNT could be either metallic or semiconducting. The higher response of SWNT sensor could be attributed to high number of semiconducting tubes, as shown by [42].

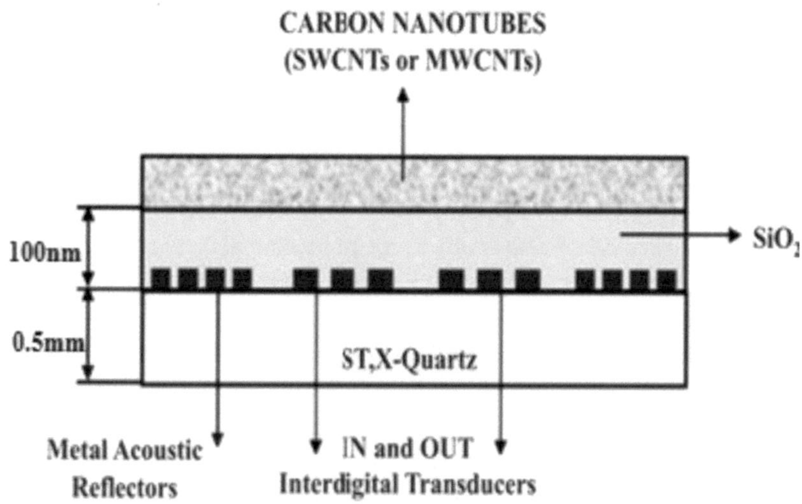

Figure 4. SAW gas sensor based on carbon nanotubes [43].

The response of the sensor was monitored as a change in resonant frequency of the SAW resonator. It was observed that there is a downshift in frequency for SAW sensors coated with MWCNT/SWCNT dispersed in toluene toward ethanol, while an upshift in frequency was recorded for SAW sensors coated with MWCNT/SWCNT dispersed in ethanol toward ethyl acetate. This increase/decrease in resonant frequency is as a result of the mass loading effect due to the molecules adsorbed on the surface of the nanotubes. This effect is manifested as perturbations in the velocity of propagation. The change in resonant frequency is directly proportional to the mass of gas that has been adsorbed.

Similarly, Cava et al. [44] employed the use of MWCNT film prepared by self-assembly technique as a sensing layer for oxygen sensor. Oxygen sensitivity was observed when the sensors were exposed to 10% oxygen in nitrogen at a temperature of 160°C. It was observed that sensors prepared by the drop-casting method were less sensitive compared with those prepared by self-assembly method. This is because the self-assembly method provides a much better distribution of the carbon nanotubes, which gives it the ability for improved gas adsorption property.

However, the limitations of using pristine CNT as a sensing material are long recovery time, low sensitivity, and sensing achieved usually at high temperature. As reported by [45], it was established that CNT-based sensors do not give response to CO and H gases at room temperature. Therefore, surface functionalization of the CNT using noble metals such as palladium or platinum is required for sensing of CO and hydrogen at room temperature [46]. Functionalization using noble metals enhances the conductivity of the CNT and thus increases sensitivity. It allows for free movement and charge transfer between the ions and the gas molecules.

Carbon nanotubes are materials which have inherent electrical and mechanical properties, and they have a hollow structure. These properties make them suitable candidates for gas sensing. Their main limitation when employed as gas-sensing materials for SAW resonators in their pristine form is that they cause short circuit of the IDTs. Therefore, to deposit the sensing layer successfully and to overcome the problem of short circuit, an insulating/guiding layer needs to be created between the IDTs and the sensing layer to form a layered structure as shown in **Figure 4**.

It has dual advantages: it serves as a guiding layer by protecting the IDTs, and it also increases the conductivity. Silicon oxide is also another type of insulating layer and has been employed for gas-sensing applications. As reported by Penza et al. [41], a two-port SAW resonator was developed for volatile organic compounds (VOCs) sensing, to deposit the sensing layer which is composed of single-walled and multi-walled carbon nanotubes; an SiO_2 insulating layer was sputtered so as to function as a guiding layer for SAW and protection of the metallic coating. Carbon nanotubes were then sprayed with an airbrush. Response of carbon nanotubes toward VOCs was then observed.

Similarly, the same authors [43] developed a SAW two-port resonator for the detection of VOCs using similar structure, with SiO_2 as an insulating layer of thickness 100 nm. Some researches employ the use of layered structure for sensing layer. The layered structure has dual function: to serve as a guiding layer and to increase the sensitivity. This could be seen in [47] was fabricated using $ZnO/LiTaO_3$ substrate and the ZnO serves as the guiding layer for the electrodes and also to enhance the sensitivity. Single-walled carbon nanotubes were then deposited using the Langmuir–Blodgett technique, and then tested toward hydrogen, ammonia, and nitrogen oxide gases, respectively.

A double surface acoustic wave resonator system was proposed by [48], as shown in **Figure 5**. In this system, the sensing layer will be fabricated and integrated separately from the SAW resonators. This will eliminate the use of guiding layer and the short circuiting of the IDTs as a result of deposition of CNT. The system was successfully used for the detection of hydrogen from 1 to 2% with functionalized CNT/Pani sensing layer.

http://dx.doi.org/10.5772/63287

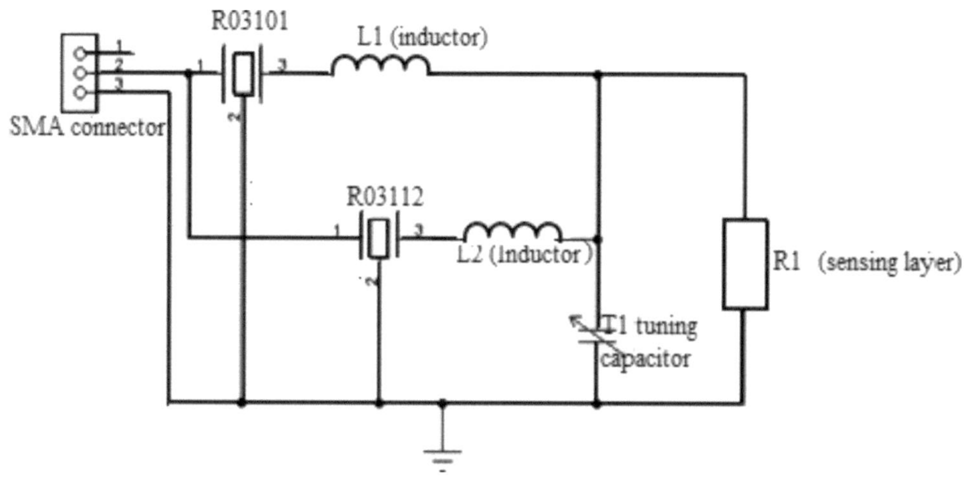

Figure 5. DSAWR system for gas-sensing applications [48].

Recently, Arab et al. [49] developed a SAW resonator gas sensor for acetone detection using MWCNT/cerium oxide nanocomposites. Preliminary results using pristine MWCNT caused short circuit to the IDTs, which is due to their high metallic behavior, and such MWCNTs were mixed with cerium oxide so as to reduce this metallic behavior and prevent the IDTs from short circuit. This gave good response to acetone vapor and ethanol at room temperature.

1.1.3.2. Gas sensors based on CNTs/nanocomposites and polymers

Conducting polymers are among the organic polymers that are suitable and promising materials for gas-sensing applications as could be recalled from Section 1. To enhance their sensitivity, conducting polymers are combined with CNTs because of their instability when used as gas-sensing material alone. However, due to the inability of sensing CO gas at room temperature using SWNT but only at elevated temperatures [50]. However, Choi et al. investigated the electrical properties of PANI/SWNT as a composite sensing material for the detection of CO and ammonia. Results obtained showed that the composite material was more responsive toward ammonia than CO gas with increased conductance for CO but decreased conductance toward ammonia [38] .

Similar to polymers, nanocomposites are also another group of promising materials that could be used to enhance gas-sensing performance. This is due to their high surface area, chemical, and mechanical properties. Nanocomposites could be integrated with metal oxides, polymers, and carbon nanotubes for improved performance. The response of polymers and CNT composites has also shown improved performance. It was established that the CNTs–polymer composites could be used to improve the mechanical properties and stabilization [45]. This is because when the CNT is integrated inside the polymer matrix, more mechanical support is provided in the arrangement of the polymer chain structure.

Sayago et al. [51] investigated the sensing behavior of hybrid polymers when used as a sensing material. A SAW gas sensor was fabricated using polyepichlorohydrin (PECH) and polyether urethane (PEUT) with different ratios of MWCNTs. Solutions of the nanocomposites were prepared with different weights of the MWCNTs and polymers, and ultrasonicated. The solutions were deposited using airbrush spraying with a metallic mask on the sensitive layer only. The sensor was tested toward volatile organic compounds, such as octane and toluene, and also toward different gases including hydrogen, ammonia, nitrogen dioxide, and carbon monoxide. Results obtained showed good response toward octane and toluene, but there were no responses to any of the gases. This means that the composite materials do not have good selectivity toward the gases but rather toward volatile organic compounds only.

Similarly, nanocomposites based on CNT and metal oxides can also be promising sensing materials. David et al. [49] employed the use of MWCNT/Ceria nanocomposites as sensitive layers for SAW gas sensor. The sensor was tested toward acetone and ethanol vapors. Frequency shifts obtained showed higher response toward acetone vapor with a frequency shift of 200 kHz at room temperature.. Similarly, nonconducting polymers could also be another class of polymers that could be used as nanocomposite materials for gas sensing. Chee Song et al. [50] fabricated a SAW gas sensor polyvinylpyrrolidone (PVP)/MWCNT nanofiber composites for hydrogen detection. The nanofibers were synthesized by electrospinning the composite solution deposited directly on the active layer of the SAW. Frequency shifts were observed with 530 Hz for 1% hydrogen concentration and 11.322 kHz of 0.25% hydrogen concentration for PVP concentrations of 1.525 and 1.025 g, respectively. This shows that the sensor with a low PVP concentration gave a higher frequency shift.

Recently, it was established that carbon nanotubes with metal nanoparticles could also be used as promising sensing materials. Mohsen Asad et al. [51] fabricated a SAW gas sensor based on thin film of single-walled carbon nanotubes decorated with copper nanoparticles as a sensing layer, which was deposited using drop-casting method. The sensor was tested toward different gases, and the effect of temperature on the sensor response was investigated. The SAW sensor was tested toward hydrogen, acetone, ethanol, and hydrogen sulfide, so as to investigate its selectivity. The highest selectivity was observed toward hydrogen sulfide gas.

1.1.3.3. Gas sensors based on graphene

Graphene is a two-dimensional form of carbon that possesses unique properties. Due to its low resistance and large surface area to volume ratio, it has shown to be a promising sensing material. In recent years, graphene and its derivatives such as pristine graphene, graphene oxide, and reduced graphene oxide (RGO) have been reported for different gas-sensing applications in the detection of different gases, including H_2O, NO_2, CO, and NH_3. The sensor response is measured as a change in the resistance of the graphene film. However, graphene has not been popularly employed in SAW gas sensors.

Guo [53] developed a SAW humidity gas sensor that is based on graphene. It was reported that for low relative humidity of less than 50%, a downshift of 1.38 kHz was observed, which was as a result of mass loading effect due to water adsorption by the graphene surface. Mass loading effect was observed to be less effective at high temperatures.

Thomas et al. [54] also developed a low-cost Rayleigh SAW resonator coated with a layer of graphene. Exfoliated reduced graphene oxide (RGO) dots were deposited in the sensing region, as shown in **Figure 6**. The sensor was deployed for the detection of different concentrations of NO_2 in air, and it exhibited good sensitivity of 25 Hz/ppm at low power and cost.

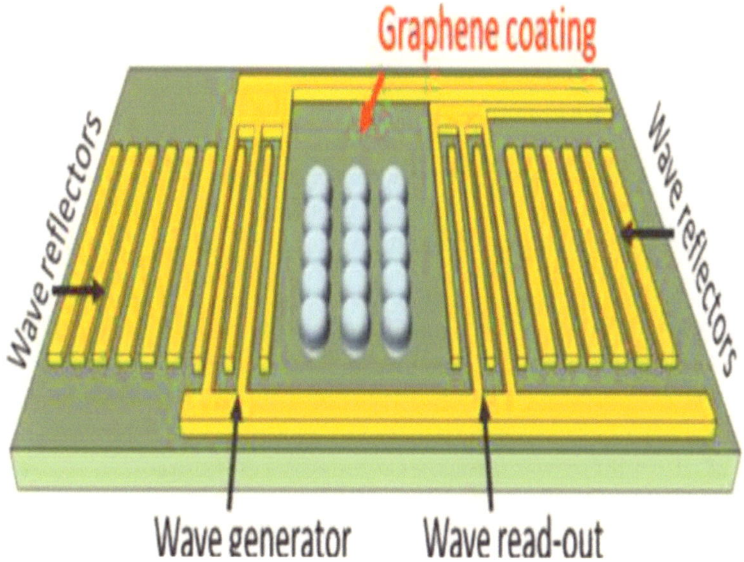

Figure 6. SAW resonator coated with graphene for NO_2 detection [54].

Similarly, Arsat et al. [55] reported SAW gas sensor based on graphene-like nanosheets deposited on $LiTaO_3$. Graphene was synthesized by reduction of graphite oxide and deployed for detection of H_2 and CO at room temperature and 40°C. A frequency downshift was observed with CO and an upshift with H_2. This was speculated to be as a result of incomplete reduction of GO with hydrazine.

1.2. Conclusion and Outlook for Researchers

SAW-based gas sensors have proved to be reliable and promising in terms in sensitivity, selectivity, high response, and recovery times. Increased sensitivity and selectivity are among the key parameters a designer requires during the development of the sensor. As the sensing layer is an integral part of the sensor, therefore knowledge of an appropriate sensing material is vital for researchers. Different sensing materials which include metal oxides, polymers, carbon nanostructures (carbon nanotubes and graphene), with their composites, have been reviewed in this article. One of the limitations of polymers when employed as gas-sensing materials is that after repeated measurements, there tends to be humidity interference. Therefore, researchers should devise a way of regeneration of the polymer, so as to maintain reproducibility during measurement. Similarly, pristine carbon nanotubes, due to their metallic nature, have limitations when employed as sensing material, because they cause short circuit of the IDTs. For sensing applications that require CNTs, a guiding layer needs to be fabricated which makes the design more costly.

Due to the interesting features of graphene, which include low cost, high surface area to volume ratio, and ease of processing, it has been shown to be a promising sensing material. However, not much literature has been reported for graphene in SAW-based gas sensors. Therefore, researchers could explore more applications of graphene as sensing material for SAW gas sensors. Graphene and its oxides are promising sensing materials due to production of less expensive and highly sensitive gas sensors. Therefore, future works should focus on employing hybrids of graphene, including graphene oxides such as ZnO, WO_3, SnO_2, CO_3O_4, etc., as gas-sensing materials. Similarly, single-layer graphene and few-layer graphene have good mechanical strength and high temperature tolerance, which makes them good candidates for sensing materials, based on literature that they have not been exclusively used for SAW gas sensors. It is therefore recommended that future works should incorporate them.

Different methods have been used for synthesis of graphene, including epitaxial growth, unzipping of CNT, mechanical exfoliation, and chemical vapor deposition, among others. Graphene produced from each technique tend to have different properties. It was shown by Saleh-Khojin et al. [56] that sensors made of CVD-grown graphene nanoribbon exhibit much higher sensitivity to gases than defect-free graphene that is produced by Scotch tape method, which is due to adsorption of gas molecules at the edges of the graphene surface. GNRs have been synthesized using CVD by [57]; therefore, work is ongoing on the application of these GNRs in gas sensing.

Author details

Mohd Nizar Hamidon[1*] and Zainab Yunusa[2]

*Address all correspondence to: mnh@upm.edu.my

1 Faculty of Engineering, Department of Electronic Engineering, Universiti Putra Malaysia, Serdang, Selangor, Malaysia

2 Faculty of Engineering, Department of Electrical Engineering, Bayero University KanoKano, Nigeria

References

[1] M. N. Hamidon, "Fabrication of High Temperature Surface Acoustic Wave Devices for Sensor Applications," University of Southampton, 2005.

[2] A. Mujahid and F. L. Dickert, "SAW and Functional Polymers," 2013.

[3] B. Drafts, "Acoustic wave technology sensors," *IEEE Trans. Microw. Theory Tech.*, vol. 49, pp. 795–802, 2001.

[4] N. Barsan, D. Koziej, and U. Weimar, "Metal oxide-based gas sensor research: how to?" *Sens. Actuators B Chem.*, vol. 121, no. 1, pp. 18–35, 2007.

[5] T. Hübert, L. Boon-Brett, G. Black, and U. Banach, "Hydrogen sensors—a review," *Sens. Actuators B Chem.*, vol. 157, no. 2, pp. 329–352, 2011.

[6] C. Wang, Y. Wang, S. Zhang, L. Fan, and X. Shui, "Characteristics of SAW hydrogen sensors based on InO_x /128° YX-$LiNbO_3$ structures at room temperature," vol. 173, no. 2, pp. 710–715, 2012.

[7] V. B. Raj, H. Singh, A. T. Nimal, M. Tomar, M. U. Sharma, and V. Gupta, "Effect of metal oxide sensing layers on the distinct detection of ammonia using surface acoustic wave (SAW) sensors," *Sens. Actuators B Chem.*, vol. 187, pp. 563–573, 2013.

[8] D.-T. Phan and G.-S. Chung, "Surface acoustic wave hydrogen sensors based on ZnO nanoparticles incorporated with a Pt catalyst," *Sens. Actuators B Chem.*, vol. 161, no. 1, pp. 341–348, 2012.

[9] A. Z. Sadek, W. Wlodarski, K. Shin, R. B. Kaner, and K. Kalantar-zadeh, "A polyaniline/WO_3 nanofiber composite-based ZnO/64° YX $LiNbO_3$ SAW hydrogen gas sensor," *Synth. Met.*, vol. 158, pp. 29–32, 2008.

[10] Y.-L. Tang, Z.-J. Li, J.-Y. Ma, H.-Q. Su, Y.-J. Guo, L. Wang, B. Du, J.-J. Chen, W. Zhou, Q.-K. Yu, and X.-T. Zu, "Highly sensitive room-temperature surface acoustic wave (SAW) ammonia sensors based on CO_3O_4/SiO_2 composite films," *J. Hazard. Mater.*, vol. 280C, pp. 127–133, 2014.

[11] Y. L. Tang, Z. J. Li, J. Y. Ma, Y. J. Guo, Y. Q. Fu, and X. T. Zu, "Ammonia gas sensors based on ZnO/SiO_2 bi-layer nanofilms on ST-cut quartz surface acoustic wave devices," *Sens. Actuators B Chem.*, vol. 201, pp. 114–121, 2014.

[12] A. Z. Sadek, W. Wlodarski, Y. X. Li, W. Yu, X. Li, X. Yu, and K. Kalantar-zadeh, "A ZnO nanorod based layered ZnO/64° YX $LiNbO_3$ SAW hydrogen gas sensor," vol. 515, pp. 8705–8708, 2007.

[13] J. L. H. T Wang, B. S Kang, F. Ren, L. C Tien, P. W Sadik, D. P Norton, and S. J Pearton, "Detection of hydrogen at room temperature with catalyst-coated multiple ZnO nanorods," *Appl. Phys.*, vol. 81, pp. 1117–1119, 2005.

[14] W. P. Jakubik, "Surface acoustic wave-based gas sensors," *Thin Solid Films*, vol. 520, no. 3, pp. 986–993, 2011.

[15] M. Penza, M. Tagliente, L. Mirenghi, C. Gerardi, C. Martucci, and G. Cassano, "Tungsten trioxide (WO_3) sputtered thin films for a NO_x gas sensor," *Sens. Actuators B Chem.*, vol. 50. pp. 9–18, 1998.

[16] I. Jiménez, J. Arbiol, G. Dezanneau, A. Cornet, and J. R. Morante, "Crystalline structure, defects and gas sensor response to NO_2 and H_2S of tungsten trioxide nanopowders," *Sens. Actuators B Chem.*, 2003, vol. 93, pp. 475–485.

[17] H. Nakagawa, N. Yamamoto, S. Okazaki, T. Chinzei, and S. Asakura, "A room-temperature operated hydrogen leak sensor," *Sens. Actuators B Chem.*, 2003, vol. 93, pp. 468–474.

[18] S. Okazaki, H. Nakagawa, S. Asakura, Y. Tomiuchi, N. Tsuji, H. Murayama, and M. Washiya, "Sensing characteristics of an optical fiber sensor for hydrogen leak," *Sens. Actuators B Chem.*, vol. 93, no. 1–3, pp. 142–147, 2003.

[19] S. J. Ippolito, S. Kandasamy, K. Kalantar-zadeh, and W. Wlodarski, "Layered SAW hydrogen sensor with modified tungsten trioxide selective layer," vol. 108, pp. 553–557, 2005.

[20] J. Kukkola, M. Mohl, A.-R. Leino, J. Mäklin, N. Halonen, A. Shchukarev, Z. Konya, H. Jantunen, and K. Kordas, "Room temperature hydrogen sensors based on metal decorated WO_3 nanowires," *Sens. Actuators B Chem.*, vol. 186, pp. 90–95, 2013.

[21] N. Van Hieu, L. T. B. Thuy, and N. D. Chien, "Highly sensitive thin film NH_3 gas sensor operating at room temperature based on SnO_2/MWCNTs composite," *Sens. Actuators B Chem.*, vol. 129, no. 2, pp. 888–895, 2008.

[22] A. Z. Sadek, W. Wlodarski, K. Shin, and R. B. Kaner, "A layered SAW gas sensor based on a polyaniline/In_2O_3 nanofiber composite," vol. 17, no. 2, pp. 4488–4492, 2007.

[23] S. J. Ippolito, S. Kandasamy, K. Kalantar-Zadeh, and W. Wlodarski, "Layered SAW hydrogen sensor with modified tungsten trioxide selective layer," *Sens. Actuators B Chem.*, vol. 108, pp. 553–557, 2005.

[24] K. Kalantar-Zadeh, D. A. Powell, W. Wlodarski, S. Ippolito, and K. Galatsis, "Comparison of layered based SAW sensors," *Sens. Actuators B Chem.*, vol. 91, pp. 303–308, 2003.

[25] S. J. Ippolito, A. Ponzoni, K. Kalantar-zadeh, and W. Wlodarski, "Layered WO_3/ZnO/36° $LiTaO_3$ SAW gas sensor sensitive towards ethanol vapour and humidity," vol. 117, pp. 442–450, 2006.

[26] G. C. Frye and S. J. Martin, "Velocity and attenuation effects in acoustic wave chemical sensors," *IEEE Ultrason. Symp.*, pp. 379–384, 1993.

[27] S. J. Martin, G. C. Frye, J. J. Spares, and A. Butler, "Gas sensing with acoustic devices," *IEEE Ultrason. Symp.*, pp. 423–434, 1996.

[28] O. G. Ricco AJ, Crooks RM, "Surface acoustic wave chemical sensor arrays new chemically sensitive interfaces combined with novel cluster analysis to detect volatile organic compounds and mixtures," *Accounts Chem. Res.*, vol. 31, no. 5, 1998.

[29] L. Al-Mashat, H. D. Tran, W. Wlodarski, R. B. Kaner, and K. Kalantar-Zadeh, "Polypyrrole nanofiber surface acoustic wave gas sensors," *Sens. Actuators B Chem.*, vol. 134, pp. 826–831, 2008.

[30] S. Kochowski, J. Bodzenta, W. P. Jakubik, and M. W. Urban, "Bilayer structure for hydrogen detection in a surface acoustic wave sensor system," vol. 82, pp. 265–271, 2002.

[31] L. Al-Mashat, H. D. Tran, R. B. Kaner, R. Arsat, K. Kalantar-Zadeh, and W. Wlodarski, "A hydrogen gas sensor fabricated from polythiopene nanofibers deposited on a 36 YX $LiTaO_3$ layered surface acoustic wave transducer," vol. 7268, no. 2, pp. 72680M–72680M-8, 2008.

[32] R. Arsat, X. F. Yu, Y. X. Li, W. Wlodarski, and K. Kalantar-Zadeh, "Hydrogen gas sensor based on highly ordered polyaniline nanofibers," vol. 137, pp. 529–532, 2009.

[33] A. Z. Sadek, W. Wlodarski, K. Shin, R. B. Kaner, and K. Kalantar-Zadeh, "A room temperature polyaniline/SnO_2 nanofiber composite based layered ZnO/64° YX $LiNbO_3$ SAW hydrogen gas sensor," pp. 208–211, 2006.

[34] A. Z. Sadek, W. Wlodarski, K. Shin, R. B. Kaner, and K. Kalantar-Zadeh, "A polyaniline/WO_3 nanofiber composite-based ZnO/64° YX $LiNbO_3$ SAW hydrogen gas sensor," *Synth. Met.*, vol. 158, pp. 29–32, 2008.

[35] T. Zhang, S. Mubeen, N. V Myung, and M. A. Deshusses, "Recent progress in carbon nanotube-based gas sensors," vol. 19, 2008.

[36] O. K. Varghese, P. D. Kichambre, D. Gong, K. G. Ong, E. C. Dickey, and C. A. Grimes, "Gas sensing characteristics of multi-wall carbon nanotubes," *Sens. Actuators B Chem.*, vol. 81. pp. 32–41, 2001.

[37] W. S. Cho, S. Il Moon, K. K. Paek, Y. H. Lee, J. H. Park, and B. K. Ju, "Patterned multiwall carbon nanotube films as materials of NO_2 gas sensors," *Sens. Actuators B Chem.*, vol. 119, pp. 180–185, 2006.

[38] L. Zhao, M. Choi, H.-S. Kim, and S.-H. Hong, "The effect of multiwalled carbon nanotube doping on the CO gas sensitivity of SnO_2-based nanomaterials," *Nanotechnology*, pp. 445501, 2007.

[39] A. H. Jayatissa, S. Member, and K. Guo, "Multiwalled carbon nanotube-based hydrogen gas sensors," pp. 0–4, 2008.

[40] M. Kumar and Y. Ando, "Chemical vapor deposition of carbon nanotubes: a review on growth mechanism and mass production," *J. Nanosci. Nanotechnol.*, vol. 10, no. 6, pp. 3739–3758, 2010.

[41] M. Penza, F. Antolini, and M. V. Antisari, "Carbon nanotubes as SAW chemical sensors materials," vol. 100, pp. 47–59, 2004.

[42] J. Suehiro, H. Imakiire, S. I. Hidaka, W. Ding, G. Zhou, K. Imasaka, and M. Hara, "Schottky-type response of carbon nanotube NO_2 gas sensor fabricated onto aluminum electrodes by dielectrophoresis," *Sens. Actuators B Chem.*, vol. 114, pp. 943–949, 2006.

[43] M. Penza, F. Antolini, and M. Vittori-Antisari, "Carbon nanotubes-based surface acoustic waves oscillating sensor for vapour detection," *Thin Solid Films*, vol. 472, pp. 246–252, 2005.

[44] C. E. Cava, R. V. Salvatierra, D. C. B. Alves, A. S. Ferlauto, A. J. G. Zarbin, and L. S. Roman, "Self-assembled films of multi-wall carbon nanotubes used in gas sensors to increase the sensitivity limit for oxygen detection," *Carbon N. Y.*, vol. 50, pp. 1953–1958, 2012.

[45] W.-D. Zhang and W.-H. Zhang, "Carbon nanotubes as active components for gas sensors," *J. Sensors*, vol. 2009, pp. 1–16, 2009.

[46] M. K. Kumar and S. Ramaprabhu, "Nanostructured Pt functionlized multiwalled carbon nanotube based hydrogen sensor," *J. Phys. Chem. B*, vol. 110, pp. 11291–11298, 2006.

[47] M. Penza, P. Aversa, G. Cassano, W. Wlodarski, and K. Kalantar-Zadeh, "Layered SAW gas sensor with single-walled carbon nanotube-based nanocomposite coating," *Sens. Actuators B Chem.*, vol. 127, pp. 168–178, 2007.

[48] Z. Yunusa, M. N. Hamidon, A. Ismail, M. M. Isa, M. H. Yaacob, S. Rahmanian, S. A. Ibrahim, and A. A. A. Shabaneh, "Development of a hydrogen gas sensor using a double saw resonator system at room temperature," *Sensors MDPI*, vol. 15, pp. 4749–4765, 2015.

[49] M. David, M. Arab, C. Martino, L. Delmas, F. Guinneton, and J. Gavarri, "Carbon nanotubes/ceria composite layers deposited on surface acoustic wave devices for gas detection at room temperature," *Thin Solid Films*, vol. 520, no. 14, pp. 4786–4791, 2012.

[50] I. Sayago, M. J. Fernández, J. L. Fontecha, M. C. Horrillo, C. Vera, I. Obieta, and I. Bustero, "Surface acoustic wave gas sensors based on polyisobutylene and carbon nanotube composites," *Sens. Actuators B Chem.*, vol. 156, no. 1, pp. 1–5, 2011.

[51] I. Sayago, M. J. Fernández, J. L. Fontecha, M. C. Horrillo, C. Vera, I. Obieta, and I. Bustero, "New sensitive layers for surface acoustic wave gas sensors based on polymer and carbon nanotube composites," *Sens. Actuators B Chem.*, vol. 175, pp. 67–72, 2012.

[52] M. Asad and M. H. Sheikhi, "Surface acoustic wave based H2S gas sensors incorporating sensitive layers of single wall carbon nanotubes decorated with Cu nanoparticles," *Sens. Actuators B Chem.*, vol. 198, pp. 134–141, 2014.

[53] Y. J. Guo, J. Zhang, C. Zhao, P. A. Hu, X. T. Zu, and Y. Q. Fu, "Graphene/$LiNbO_3$ surface acoustic wave device based relative humidity sensor," *Optik (Stuttg).*, vol. 125, pp. 5800–5802, 2014.

[54] S. Thomas, M. Cole, A. De Luca, F. Torrisi, A. C. Ferrari, F. Udrea, and J. W. Gardner, "Graphene-coated Rayleigh SAW resonators for NO_2 detection," *Procedia Eng.*, vol. 87, no. 2, pp. 999–1002, 2014.

[55] R. Arsat, M. Breedon, M. Shafiei, P. G. Spizziri, S. Gilje, R. B. Kaner, K. Kalantar-zadeh, and W. Wlodarski, "Graphene-like nano-sheets for surface acoustic wave gas sensor applications," *Chem. Phys. Lett.*, vol. 467, pp. 344–347, 2009.

[56] S. Gupta, S. Chatterjee, A. K. Ray, and A. K. Chakraborty, "Graphene–metal oxide nanohybrids for toxic gas sensor: a review," *Sens. Actuators B Chem.*, vol. 221, no. 2, pp. 1170–1181, 2015.

[57] Z. Yunusa, S. A. Rashid, M. N. Hamidon, S. Hafiz, I. Ismail, and S. Rahmanian, "Synthesis of Y-tip graphitic nanoribbons from alcohol catalytic chemical vapor deposition on piezoelectric substrate," *J. Nanomater. Hindawi*, vol. 2015, pp. 2–9, 2015.